A PLUME BOOK

EUCLID IN THE RAINFOREST

JOSEPH MAZUR is professor of mathematics at Marlboro College, where he has taught a wide range of classes in all areas of mathematics, its history, and its philosophy. He lives with his wife in Vermont.

"Joseph Mazur brilliantly explores the symbiotic relationship between the physical and the mathematical worlds. He asks the questions: How do we know that the world is what we experience it to be? Can logic guide us through the rainforest of science and math and provide us with a chance to discover the underlying foundations for their truths? In his highly original search, Mazur is a brilliant forester whose graceful pursuit leads him to understand the logical bases of human reason. Mazur has given us a stylish and seductive book that convinces the mind even as it delights the soul."
　　—PEN American Center (finalist, 2005 Martha Albrand
　　　Award for First Nonfiction)

"This charming book radiates love of mathematics . . . and of life. Mazur weaves elementary explanations of a wide range of essential mathematical ideas into narratives of his far-ranging travels. . . . This book is a treasure of human experience and intellectual excitement."
　　—*Choice* (2005 Outstanding Academic Title)

"A rare example of a universally appealing math book."
　　—*Mathematics Teacher*

Euclid in the

Rainforest

Discovering Universal Truth in Logic and Math

~

Joseph Mazur

A PLUME BOOK

PLUME
Published by Penguin Group
Penguin Group (USA) Inc., 375 Hudson Street, New York, New York 10014, U.S.A.
Penguin Group (Canada), 90 Eglinton Avenue East, Suite 700, Toronto,
Ontario M4P 2Y3, Canada (a division of Pearson Penguin Canada Inc.)
Penguin Books Ltd., 80 Strand, London WC2R 0RL, England
Penguin Ireland, 25 St. Stephen's Green, Dublin 2, Ireland
(a division of Penguin Books Ltd.)
Penguin Group (Australia), 250 Camberwell Road, Camberwell,
Victoria 3124, Australia (a division of Pearson Australia Group Pty. Ltd.)
Penguin Books India Pvt. Ltd., 11 Community Centre,
Panchsheel Park, New Delhi – 110 017, India
Penguin Books (NZ), cnr Airborne and Rosedale Roads,
Albany, Auckland 1310, New Zealand (a division of Pearson New Zealand Ltd.)
Penguin Books (South Africa) (Pty.) Ltd., 24 Sturdee Avenue,
Rosebank, Johannesburg 2196, South Africa

Penguin Books Ltd., Registered Offices: 80 Strand, London WC2R 0RL, England

Published by Plume, a member of Penguin Group (USA) Inc. Previously
published in a Pi Press edition.

First Plume Printing, August 2006
10 9 8 7 6 5 4 3 2

Copyright © Joseph Mazur, 2005
All rights reserved

CIP data is available
ISBN 0-13-147994-6 (hc.)
ISBN 0-452-28783-9 (pbk.)

Printed in the United States of America

In memory of David Fowler, who, with humor and scholarly insight, introduced the world to refreshing new ways of appreciating Euclid and the mathematics of Plato's Academy

Contents

Introduction

I came to understand mathematics by way of a Russian novel. On the morning of my seventeenth birthday, my older brother presented me with two books that I, as he put it, "might enjoy reading." One was a 532-page paperback of Dostoyevsky's *Crime and Punishment*; the other was a 472-page algebra textbook. My brother, having the gift of quick mathematical insight, hadn't noticed that one does not read a textbook of modern algebra in the same way that one reads a Russian novel. The next morning, I began reading the novel without ever getting out of bed and finished sometime late that night, having skipped lunch and supper. Raskolnikov, after swinging the axe with both arms and bringing the blunt side down on the old woman's head, made me feverish and deliriously spellbound. On reaching the end of the novel, trust in my brother's choice of mesmeric literature was so strong that I turned to my other book, expecting it to be as riveting as the first. The next morning, I pondered a sentence on the second page for hours: "Clearly, any result which can be proved or deduced from the postulates defining an integral domain will be true in any particular integral domain...."

Stuck at *clearly*, I dressed and once again spent the day attempting to understand my new book, but this time I did not get past page five. That summer, I struggled to get past the first few chapters. What was modern algebra about? By the fourth chapter, I was well into abstractions, having completed as many exercises as I could. But I could not understand the relevance of all these mathematical abstractions to life itself.

The joys of feeling confidence in solving problems and the emotion of witnessing beauty at the end of a proof are intense. Mathematics gradually became for me as a teenager a range of mountains to climb. Challenges of ascending through the thin air of abstraction only made vistas from the summits more magnificent. With a firm proven foothold on one peak, I could see others beckoning me higher from above misty clouds covering valley paths through rainforests of flowering ideas.

In thirty years of teaching mathematics, I have collected stories of extraordinary students and fellow mathematicians trying to reach peaks from the steepest faces. Their stories are about the climb, the view from even the smallest peak, the excitement of discovery, the investigation of unknown intellectual encounters with beauty, and the confidence of feeling certain about mathematical proof. They are human stories, ultimately not so removed from the excitement of a Russian novel. I begin to see my brother's inadvertent point.

But this book also has another point to make about mathematics, about logic, about scientific truth. To appreciate modern mathematics, we inevitably must examine how mathematics is communicated, question what makes us *feel* persuaded by proofs of theorems. What is proof? It might seem odd to find that, even in mathematics, a subject envied for its precision, there is no universally accepted answer. A formal answer might be that it is an ordered list of statements ascending from an established fact (axiom, theorem and so on), each statement logically derived from the one preceding. However, mathematicians follow a more informal practice. Many theorems accepted and used in mainstream mathematics have proofs that hardly conform to any rigorous definition of proof.

Mathematics enjoys a reputation for being an intellectual pursuit that generates universal *truths*. But contrary to what many of us think, those truths are not communicated through airtight chains of logical arguments. The essence of proof contains something more than just pure logic, just as music is more than just musical notes. It might seem strange to think that, even though mathematics seems to be independent of culture, opinion plays a central role in the profession. How do mathematicians know when a proof is complete? Is it complete if nobody can find an error? Or does it come from an inner feeling that plays with opinion through knowledge and experience?

A significant part of the feeling of being mathematically right comes from experience with logic developed by the practice of rational criticism and debate. Sometime early in the sixth century B.C., two things happened to dramatically alter the way Western civilization explained the world. The first was the use of cause and effect, as opposed to the supernatural in explaining natural phenomena; we might say that nature was first discovered then. The second was the practice of rational criticism and debate. These fresh developments occurred after a time of great political upheaval in the eastern Mediterranean, which led to profound changes in the political structures of Greek cities. Democracy in Athens meant that citizens could participate in government and law, freely debating and questioning political ideas. Before the establishment of the Greek city-state, a change in rule usually meant merely a change from one tyrant to another. Greek philosophy, according to tradition, began in 585 B.C. when Thales and other Ionian merchants traveled to Egypt and other parts of the known world. They returned rich with information about applications of mathematics related to building practices. One can imagine Thales thinking and analyzing the essence of what he had learned during those long voyages home as his ship crossed the Mediterranean and sailed up the Aegean coast back to Miletus, his hometown on the coast of modern Turkey.

For the next three hundred years, from the time of Thales and Pythagoras, the founders of Greek philosophy, through Plato and his school in Athens, to Euclid and the founding of the Museum in Alexandria, logical reasoning developed into a system of principles empowering investigations of the purely abstract immaterial world of mathematics. A third defining moment came shortly before Euclid wrote his *Elements* when Aristotle formalized ordinary logic. He constructed fourteen elementary models of logic, such as "All men are mortal; all heroes are men; therefore, all heroes are mortal." By 300 B.C., the thirteen parchment rolls of Euclid's *Elements* were written and logical reasoning had matured enough to be reduced to a handful of rules. Part 1 of this book, "Logic," is about this kind of logic.

But logical reasoning could not address the weirdness of infinity. An incredulous Greek mathematician named Zeno constructed polemics against motion, which used supposedly iron chains of logic to tangle arguments into clanging self-contradictions. Plato tells us that Zeno came to Athens from Elea (on the west coast of Italy) with his lover, Parmenides, for the Panathenaea Festival. While there, presumably between events, Zeno read from his works to a very young Socrates. According to one of Zeno's many arguments, even the swift-footed Achilles could not overtake a slow-crawling tortoise if the tortoise was given any head start. This, Zeno argued, is because the moment Achilles reaches the tortoise's starting point, the tortoise would have moved to a spot farther ahead; at that point, the argument repeats, with the tortoise being given a new head start. Achilles would have to repeat this forever just to catch up with the tortoise. In another argument, Zeno shows that movement is impossible because, for a body to move any distance, it must first get to half the distance, then half the remaining distance, and so on, forever having to get to half of some remaining distance and, hence, never reaching the full distance.

Zeno might have raised these puzzles to provoke intellectual discourse or merely to irritate Athenian philosopher/sports fans. He was known as "the two-tongued Zeno" because he often argued both sides of his own arguments, which usually involved either the infinite or the infinitesimal and had an enduring effect on the development of geometry. It took time. Except for Zeno and Archimedes's brief noble attempts at understanding infinity shortly after Euclid, direct confrontation with the infinite had to wait almost two thousand years until tradition-bound rules of logical reasoning were relaxed to address some of the difficulties Zeno raised. In 1629, Bonaventura Cavalieri, a student of Galileo, devised a scheme for sidestepping the issues raised by Zeno, deliberately ignoring problems with the logic of his own arguments; oddly, his arguments led to correct results. Cavalieri's great contribution was to let intuition, rather than logic, guide mathematics. His ideas created the driving force behind the invention of calculus, the new math responsible for fantastic applications to the real world, from predicting planetary motion to the design of musical instruments.

Cavalieri's methods relied heavily on strong intuition. For almost two hundred years, new mathematical concepts, those that outgrew the bounds of ordinary logic, were guided and accepted by intuition rather than by logic. Strong intuition carried mathematics to new and glorious heights until things began to go wrong in the eighteenth century, when inconsistencies began to sprout. By the middle of the nineteenth century, intuition and logic were at loggerheads. Theorems that were once proven by intuition were being proven false by logic. A new kind of logic was needed, one capable of working with the rich intricacies of the infinite and infinitesimal. That new logic had to wait until the late nineteenth century for the discovery of *set theory*, the branch of mathematics that deals with the proper way to define numbers. Set theory gives us the axioms of arithmetic and leads to deep questions concerning the foundations of mathematics itself. Set theory presented us with a general unifying language for all branches of mathematics.

Georg Cantor had the appealing title of Extraordinary Professor of Mathematics (at the University of Halle in Germany). He developed set theory in the nineteenth century to study the real numbers and, by doing so, was led to one of the most revolutionary results in mathematics: that there are different *sizes* of infinity. What kind of logic led to that notion? Cantor spent a great deal of time writing philosophical and theological treatises in defense of his results on the infinite because they defied intuition. At the same time, he had a passion for Elizabethan literature and spent much of his time attempting to prove that Francis Bacon wrote Shakespeare's plays. He played on the edge of logic.

The axioms of set theory were not formulated before the beginning of the twentieth century, after many mathematicians had done a great deal of work building the correct framework for their foundations. On the other hand, Shakespeare is still credited with writing his plays.

In 1931, Kurt Gödel surprised the mathematics community by showing that the axioms of set theory were incomplete; in fact, he showed that, no matter how many new axioms are added to the system, there would always be a statement that cannot be proved or disproved within the framework of the axioms of set theory. When we say *proved or disproved*, we mean that nobody in the eternal future will ever be able to prove or disprove the statement. This must have been as great a shock to mathematics as Pythagoras's discovery that one ruler cannot measure both the side and diagonal of a square. Even Zeno's paradoxes could not rival this discovery. The persistent problem is, as it always has been, how some things never seem to end. Part 2 of this book, "Infinity," is about the logic of infinity.

Though logicians have problems with the formalisms of axiomatic set theory, everyone readily acknowledges that we are able to count, do amazing mathematics, and verifiably build and support science on the shoulders of mathematics. Problems with formal logic do not seem to interfere with material reality. The cost of relaxing the requirements from airtight proof to plausible proof has a great benefit: It validates the scientific method. Sir Francis Bacon, the father of the scientific method (and not the author of Shakespeare's plays), suggested that deductive reasoning is not appropriate in investigations of the material world. He argued that one could arrive at *plausible* general conclusions by observing special concrete cases.

Science relies on three kinds of reasoning. Surely, it relies on ordinary logic and, implicitly, on the logic of infinity, but it relies most heavily on plausible reasoning. It is based on the idea that what one finds true often enough is true. In mathematical proof, *often enough* means *infinitely often*, but scientific proof is far more relaxed. The sun rose in the sky often enough in my lifetime for me to believe it will rise again tomorrow. On the other hand, though I have never experienced a devastating earthquake, *never* is not *often enough* for me to believe that I will not experience one in the future. Whereas ordinary deductive reasoning inescapably forces the specific cases from general assumptions, plausible reasoning takes the opposite path and argues from specific observations to general (but only plausible) conclusions. No one can deny that this seems to weaken the strength of truth to the status of the plausible, but when

Sir Francis Bacon introduced this notion in 1620, he changed our understanding of knowledge, and the subsequent mathematics (probability and statistics) that was built to support his idea changed science forever, although a sure mathematical footing that quantifies plausibility would have to wait another hundred and fifty years for Thomas Bayes. The story of plausible reasoning is what Part 3 of this book, "Reality," is all about.

These are the essential forms of human reasoning and logic, what we humans consider proof that something is true: ordinary logic, which is concerned with proof and classification; the logic of infinity, which is concerned with infinity and number; and plausible reasoning, which is concerned with probability and nature—even rainforests.

Math at first sight is often intimidating—not just for novices, but even for trained scientists. Unfortunately, it is not always so clear as, say, the Hindu proof of the Pythagorean theorem, which is simply a picture and the word *Behold*. But in this book, I hope to show the beauty of mathematics and the pleasures of trekking through the mathematical landscape by climbing peaks to see rainforests of ideas through slow-moving clouds, just as Euclid did twenty-three hundred years ago. And after a long journey into abstraction, it's refreshing to return to the plausible logic that rules the science of the natural world.

PART I

Logic

CHAPTER 1

The Search for Knowledge School

An Introduction to Logic and Proof

"Number theory effortlessly produces innumerable problems, which have a sweet, innocent air about them—like tempting flowers."

"Flowers?"

"Mm-hmm." He also says that, *"it swarms with bugs waiting to bite the flower lovers who, once bitten, are inspired to excesses of effort."*

—Daphne Clair quoting Barry Mazur, Summer Seduction

The mix of fragrances from blooming yucca, violets, red jasmine, and frangipani carried on the vapors of rising mists above the humid Orinoco River are intoxicating. It is the perfume of the Venezuelan rainforests, where tobacco, banana and coffee grow wild and barbets call so loudly you can hardly hear nearby rapids. I was there at a time when one could still get a doctorate in anthropology from stories of undiscovered indigenous tribes in one of the last unknown corners of the world, a time when roads connecting Venezuelan cities and towns were unpaved or didn't exist at all. Young with mettle in adventure and eager for cerebral exercise, it was there on the edges of the primal rainforest that I first stumbled over notions of logic and what it means to believe a proof.

Before I knew anything about what anthropologists were or what they did, I read Hamilton Rice's accounts of his encounters

with the Yanomami Indians living along the Orinoco River in south-
ern Venezuela and decided to travel to the remote Venezuelan village
of La Esmeralda on my own. It was 1960, several years before
Napoleon Chagnon's best-selling book *Yanomamo: The Fierce People*
was published.[1] I got the required inoculations, bought malaria pills,
and went to the Venezuelan consulate in New York to apply for a visa.

I was disappointed to learn that it would take two months to
process my visa. Thinking it would be faster to get a visa from a small
consulate near Venezuela, I flew to Aruba, which is just sixteen miles
north of the coast of Venezuela. I arrived there on a Sunday to find
that the consulate was open only on Wednesdays. I waited three days
at the edge of boredom. At 9:00 a.m. on Wednesday, I walked up the
two flights of creaking stairs in a dark, alcohol-reeking, wooden
office building. The sign outside the consul's office door clearly indicated that the office would be open from 9:00 to noon. I waited. At
about 11:15, a Sydney Greenstreet look-alike in a white suit and
Panama hat arrived smelling of rum. He was the Venezuelan consul.

"Can I help you?" he asked scornfully.

"I would like a visa for Venezuela," I said, wondering if he would
ask me why.

"Yes, I can get you a visa, but it will take two months!" he said
with a snickering smile.

Defeated by the thought that I had traveled a thousand miles to a
barren island for the same deal that I could have had in New York, I
thought of the crazy idea of getting a fisherman to bring me across
the short stretch of water separating Aruba from the mainland.
However, the consul's snicker gave me hope, so I asked if there was
any other way to speed up the process.

"Certainly!" he said. "For five dollars, you could have it today!"

I placed a five-dollar bill on his desk. "Your passport please," he
demanded. He pulled out a mechanical stamp machine from the
center drawer of his desk, found an empty page in my passport, and
stamped it with a visa to Venezuela. I took the next flight to Caracas.

An Englishman a bit older than me sat beside me on the plane.
"Have you ever been to Caracash?" he asked with a proper English
accent embroidered with an S-lisp while extending his greeting

hand. "My name is Roger Hooper," he continued with a look suggesting that I should have heard the name before. Upon learning my plans to travel to the Orinoco River, he proceeded to tell me about his dream of doing the same. He proceeded to tell me about Joseph Conrad's *Heart of Darkness*, as if I had never heard of it. I was absorbed in his story almost as much as he was until a lightning storm struck without warning. Passengers around us anxiously clutched at armrests as we bounced from air pocket to air pocket. The plane pitched forward and backward, rocked from side to side, and shook with each threatening lightning bolt as it passed through torrential rain and foreboding darkness. Roger casually went on with his story, unaware that I was no longer listening. The plane came to an abrupt stop, listing to the left while torrential rain continued to pour over the windows; I could see nothing outside.

My neighbor continued talking as if nothing had happened. A minute passed before the pilot announced that, though the plane had come to a safe landing, it would not be easy to disembark. The plane was in water up to its belly. The pilot had chosen to emergency-land his plane in the safety of water rather than risk an airport landing close to rugged mountains in a severe storm. We disembarked on a makeshift ramp constructed by curious emergency personnel. As we were coming off the plane, Roger did not say a word about the event, nor a word about the inconvenience of our waterlogged luggage. He continued to lisp about the Congo, but I was processing almost nothing of what he was saying and was beginning to be annoyed. I was going to the Orinoco River, not the Congo. He said, in his first reference to our aircraft drama, that he, too, was headed for the Amazon on an adventure far more exciting than any emergency landing.

When I finally did get around to telling him my name, he blankly announced that he knew of a mathematician by that last name and asked if there was any relation. I had grown used to saying "he's my brother" whenever people asked the same question. At that, he said that he also majored in mathematics and unleashed the grand story of Fermat and his Last Theorem, the famous theorem that says that $x^n + y^n = z^n$ has no nonzero whole-number solutions for x, y, and z

when *n* is greater than 2. He could make the Caracas airport flight schedule seem like a Noel Coward play. Roger explained the theorem with the math embroidered into a fabric of supporting characters. In fact, all we really know for sure is that when Fermat died, his son Samuel discovered a copy of a translation of Diophantus's *Arithmetica* in the library with a marginal note in his father's handwriting: "I have discovered a truly remarkable proof, which this margin is too small to contain."

Few of us are taught the rules of logic or principles of proof. Even mathematicians seem to innocently learn them indirectly through instinctive knowledge and experience of language, just as the toddler does in acquiring the elements of grammar. However, when we intellectually reflect on logic and proof, what they are and what they do, we tend to agree that, formally, a proof consists of a finite collection of well-formed statements linking back, by rules of inference, to elementary assumptions. For example, if we want to prove that the base angles of an isosceles triangle[2] are equal, we must start somewhere. Proof means that belief is universal. If Euclid's proofs are to be accepted, all parties must agree that "there is a unique line passing through any two points," which is exactly Euclid's first elementary assumption. This assumption, together with four others, forms a footing to support a superstructure of Euclid's geometry.

Margin notes were typical for Pierre Fermat. He was a lawyer and judge in Toulouse, and, though considered the greatest mathematician of his day, he did mathematics only in his spare time. He didn't publish his works, but he preferred to write them as gifted letters to friends and professional mathematicians. Pierre frequently wrote his thoughts in the margins of his books, often while listening to briefs at court. But Roger painted a fanciful picture of Samuel, his appearance as a young man with bulging eyes, an unusually long but handsome nose, a sullen face, and a tragic mouth that always seemed to be lamenting something lost. According to Roger, Samuel was asleep in his library when a howling wind blew a window open. The window frame knocked Diophantus's *Arithmetica* off a sill, face down, and opened to a page biased by a pressed dry rose. And there it rested until Samuel lit a candle to see the short comment responsible for a great deal of modern number theory.

When Roger finished his story, he suggested that we should head for La Esmeralda together. I was only eighteen and Roger was ten years older. He already had a receding hairline at the sides of his head and the beginnings of a double chin that he was trying to hide behind a newly cultured beard. And, though he was definitely peculiar, insensitive, and possibly even crazy, I wanted someone to accompany me on a journey that was beginning to frighten me. Roger had the annoying habit of cracking his knuckles at odd times when he was silent. He would dovetail fingers from both hands, turn his palms outward, and push both hands forward to the sound of multiple cracks. However, he was fluent in Spanish, and I needed a translator. I had no idea what trouble I was about to get into and naively thought I could make my way into the jungle on a few provisions, a tent, and malaria pills.

At his suggestion, we hitchhiked a ride with a Venezuelan army convoy traveling five hundred miles into the jungle along dirt roads to Cabruta on the Orinoco River. It was an innocent time, when two foreign civilians could unofficially still do such things. The convoy would go no further than Cabruta, for there was no bridge to cross the river, but we accepted the ride, thinking that at least we would get to the Orinoco, even if it were a thousand miles downstream from La Esmeralda. After about two hundred and fifty miles, we came to a washed-out mountain switchback. A canvas-covered army vehicle was tottering over a precipice and about to fall into a ravine several hundred feet below. It was nighttime and lightly raining; the bugs were out to devour anyone stupid enough to stand on the road unprotected. More than twenty soldiers took positions to push the vehicle onto a more solid road surface, and my position was at the right taillight, the main attraction for mosquitoes and diving beetles in the area. My clothes were covered with tightly gripping diving beetles. Roger was thrilled. This was the adventure he had come for.

We gave up for the night when the rain started to come down heavily. The road was impassable, so we slept huddled under the canvases of the trucks. I can't recall why the tottering vehicle was not tethered or winched to one of the other trucks, but it wasn't. In the morning, the sun rose above several isolated mists that passed with gentle breezes. A choir of birds awakened us with what seemed to be a thousand different simultaneous songs. Several of us with stiff bodies from sleeping in awkward positions rose with the sun to hear and feel the road give way, sending the tottering vehicle to its grave at the bottom of the ravine. I looked out in time to see Roger standing on the road, peering down at the fallen truck, laughing.

A cook prepared breakfast of potatoes and eggs on a propane stove, while a small group of three young, tough soldiers and a platoon sergeant argued about how to salvage the fallen vehicle. Miguel Ramos, the sergeant, was more concerned with the contents of the truck. Eating peaches from a can, he gave a stout order to salvage the contents. So several fifty-five gallon oil drums and dozens of sealed wooden crates marked in alphanumeric code were hoisted to the road, while Jesus and two other soldiers whose names I no longer recall continued to argue about how to salvage the truck itself.

The truck weighed two tons, and the winch cable capacity was only one ton. I volunteered to say that if the truck could be straightened out, it would take only about 1.4 tons of cable tension to winch the truck back to the road.[3] At the time, I thought it to simply be an academic point, since the cable's capacity was still smaller than what was needed to hoist the entire truck. But I had to convince the group that they needed a winch with only a 1.4-ton capacity.

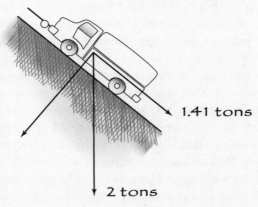

1.41 tons

2 tons

I used a stick to draw a diagram in the dirt, trying to explain the trigonometry of forces involved while a conspicuously colorful object, far off in the distance over the steep hillside of wildflowers, distracted my audience. I stopped my explanation to join the diversion.

It was a wild apricot tree two hundred feet away in the middle of the field, surrounded by tall, particularly beautiful flowers with very tall pistils and alternating red and black petals. In this wonderful field of yellow and blue flowers, that unusual red-and-black one was the only one swarmed with bees. No one had ever seen such a flower. For several minutes, we were captivated.

We approached the tree to find that what we thought was one apricot tree was actually two intertwined trees. As we stuffed our pockets with wild apricots, an idea leaped in front of me: If we could get the winch on the fallen truck to work, we could intertwine two winches to work together with a capacity of two tons—more force than necessary! The answer depended on whether the fallen truck would start. Luckily, a tree had broken the fall as the truck rolled backward, so the only damage was to the rear end. Jesus started the truck and got the winch working. The rest was easy. Entwined cables connected the fallen truck to a tethered truck on the road. The two winches synchronically rewound, slowly pulling the fallen truck up the dirt incline. A small piece of entirely theoretical trigonometry solved a thoroughly physical problem, a testament to the power of mathematics. The operation worked so smoothly that Jesus asked me to explain how I made my calculations. He was hooked.

And so was I. For the remainder of our trip, Roger enthusiastically taught us applications of trigonometry and beguiling mathematics.

Mathematics was not always what it is today. Its formal rigor and structure did not always rely on finite collections of well-formed statements linking back, by rules of inference, to elementary assumptions. No, in Thales's time (around 600 B.C., a half-century before Pythagoras), when Western mathematics inherited a fortune in concrete applications from a thousand years of Babylonian and Egyptian calculations, when abstract so-called theorems in need of proof were beginning to surface on the mathematical horizons, mathematical proof was far more relaxed and casual. Arguments were persuasive but not as rigorous as those that would come in the next three hundred years, before Euclid, when the ideas for structured proof based on elementary assumptions emerged.

One of the earliest existing histories of geometry is one that comes to us from Proclus, a philosopher and historian who summarized an earlier history by Eudemus of Rhodes. According to several historians, from Proclus to Plutarch, Thales of Miletus introduced abstract geometry, a new phenomenon that was to excite further Greek creations.[4] The Rhind papyrus, now in the British Museum, is a handbook of practical problems dating from about 1550 B.C. The papyrus calls itself "a guide to accurate reckoning of entering into things, knowledge of existing things all." Everything in the papyrus is stated in terms of specific numbers and nothing is generalized. Rectangles are measured, but only when their sizes are specified; right-angled triangles[5] mark out areas only when the sides are specified. During Thales's lifetime, Egyptian mathematics was still relying on tables to solve problems from surveying to banking, just as modern bankers rely on tables or computer programs to compute mortgage payments. Thales's revelation was that there are precise relationships between parts of figures and that such relationships enable one to find one part by knowing others.

It may seem obvious to us that the height of a right triangle, whose hypotenuse is, say, five units long, absolutely depends on the length of the base. Equally obvious is the thought that one can know what that dependency is. We know that the height is the square root of twenty-five minus the square of the length of the base. But this was not something in the Egyptian schoolboy's textbook.

Modern Western mathematics began with Thales. We are told that he is the first to "demonstrate" the following:

1. The circle is bisected by its diameter.
2. The base angles of an isosceles triangle are equal.
3. The vertical angles of two straight lines are equal.
4. The right-angled triangle can be inscribed in a circle in such a way that its hypotenuse coincides with the diameter and its right angle sits on the circle.

These are extraordinary statements about *every* circle and *every* isosceles triangle.

He must have known that the sum of the angles of a triangle is 180°, or the sum of two right angles. The persuasive argument of his time must have been something like this: Any triangle can be split into two right triangles simply by dropping a perpendicular from the largest angle to the side opposite that largest angle.

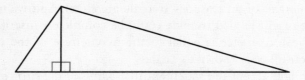

Take any one of these two right triangles, inscribe it in a circle, and connect the right angle to the center of the circle.

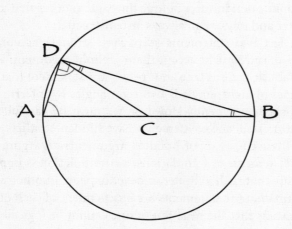

You now have two isosceles triangles. But Thales knew that the base angles of an isosceles triangle are equal. Since angle *ADB* is a right angle, we see that the sum of all three angles must be two right angles because angle *ADC* + angle *CDB* equals one right angle, and angle *DAC* + angle *DBC* = angle *ADC* + angle *CDB*.

This could be one of the earliest examples of mathematical persuasion, but not a proof in the modern sense of the word because it simply follows a picture without inferences from axioms or established statements. To make the argument into a proof, we must infer the argument from some collection of established statements that can never be disputed. What are those established statements? We used the fact that the points of the circle are equidistant from the center to know that the triangles are isosceles. We used the fact that the base angles of an isosceles triangle are equal. And we used the fact that a right-angled triangle could be completely inscribed in a circle whose diameter is equal to the hypotenuse of the triangle. These were all facts known to Thales.

We might accept the statement that *points of the circle are equidistant from its center* as part of the definition of a circle. However, the statement that *base angles of an isosceles triangle are equal* is much harder to accept. It's not obvious. It relies on an argument composed of several more elementary statements, one of which is that two points determine a unique line, and another that a circle of given center and radius can always be constructed.

These last two statements seem very strongly acceptable—so acceptable that we might accept them without argument. We could simply establish these as true and "prove" the statement that the sum of the angles of any triangle is two right angles by inferring it from the established statements. However, we must be careful. Suppose that the two established statements have hidden conflicts. What if, after one thousand years of building arguments on arguments that use these statements as foundations, a contradiction is implied? We must be sure that such a thing can never happen. In other words, we must be sure that the statements are independent of each other.

This means that the negation of one cannot be logically derived from the other. Discovering this was the great achievement of early Greek mathematicians in the three hundred year span between Thales and Euclid. When Thales generalized the practical mathematics of the Egyptians, other mathematicians made other general discoveries. The Pythagorean School discovered relationships between the sizes of the sides of right triangles. Eventually, the work turned to organizing general statements, systematizing and ordering statements in a network of inferential statements, one following from another.

Although mathematics appears to be built from a collection of self-evident assumptions, it does not develop directly from them. Thales had no established postulates, yet his arguments are as persuasive as any that could be drawn from a rigorous logical sequence of inferences. Indeed, his arguments are simply convincing and persuasive, with little room for doubt. Ideas for theorems seem to come from intuition that follows experience. Experience enough drawings of right triangles inscribed in a circle, and you will discover that the sum of the angles of a triangle is 180°. Experience drawing enough triangles, and you will discover that you can split any triangle into two right triangles. That's the way you would demonstrate a theorem: persuasion by experience. Later, when you are sure that your intuition is right because you have persuasive arguments to show that it is right, you might want to find your theorem's position in the web of others that are ultimately built on self-evident assumptions.

Jesus and I developed an insatiable appetite for mathematics. At the time, I was studying architecture and had no real training in mathematics. The solution to the truck winching problem came to me by accident, without thought. It not only impressed Jesus, but it inspired both of us to consider studying more mathematics. Jesus decided that he would go to university to study engineering.

I started thinking about proof. Pushing the limits of my knowledge at the time, I explained a simple proof of a proposition found in Book I of Euclid's *Elements*. It says that an isosceles triangle has its base angles equal. Thales proved this proposition sometime back in the fifth century B.C. Later, Pappus, a Greek mathematician who lived around A.D. 300, gave a different proof—a curious proof: Flip the given isosceles triangle around its vertical axis. The triangle is unchanged. Since the angle on the left of the base coincides with the angle on the right, the two base angles *must* be equal.

[handwritten margin note: Triangle on verticle axis = No change]

If the sides of such an isosceles triangle are extended equal distances beyond the base, the angles under the base will also be equal. This was proven by the same "flip" argument.

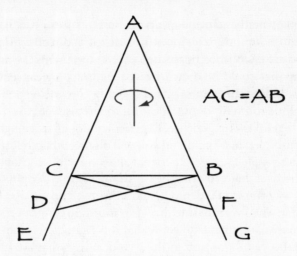

AC=AB

In medieval times, this theorem was the end of the road for courses in mathematics. It was known as "the bridge of fools" because the figure looks like the truss of a bridge and also because only a fool couldn't make it across.

This kind of heuristic proof was acceptable in Thales's time, but not in Euclid's. The same proposition appears as Proposition 5 in Book I of Euclid's *Elements*, with a proof based on far stricter standards. Although heuristics might have played a central role in guiding discovery and acceptance of the theorem, rigor would inevitably challenge any argument not based on strict rules of proof.

I confess that at the time of meeting Jesus, my knowledge of mathematics did not extend beyond this bridge. But later that bridge was moved closer to that wonderful theorem in mathematics called the Pythagorean theorem, which states that the sum of the squares of the lengths of the sides of a right triangle equals the square of the length of the hypotenuse.[6] According to the *Guinness Book of World Records*, the Pythagorean theorem has the most known proofs: 520 different proofs.[7]

When Jesus heard about Pythagoras's theorem, he tested it on all the triangles he knew, making light sketches of right triangles and measuring sides and hypotenuses. If it looked like it was true for a few triangles, he would not be satisfied until he sketched others and tested the statement again and again on squat, fat, and tall triangles.

How could he have convinced himself of the truth? He would have had to draw some fairly precise right triangles and take reasonably precise measurements of the sides and hypotenuses. Although this method might persuade some people, it wouldn't be enough to convince Jesus, and it certainly wouldn't be enough to convince mathematicians.

However, it doesn't take much evidence to form *opinions* about mathematical statements. There was good evidence to suggest that the theorem is true. So Roger gave Jesus one of the standard arguments in favor of the truth.

"Draw any right triangle and three squares, one on each side and the third on the hypotenuse," he said.

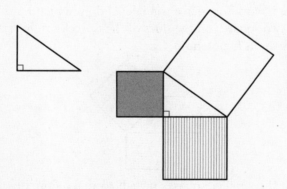

"You can place the contents of the lower square into the tilted square on the hypotenuse with room to spare. Next, you see that the remaining square on the left will fit into the remaining space of the square on the hypotenuse without room to spare. This should convince you that the sum of the areas of the squares on the sides is equal to the area of the square on the hypotenuse."

"It doesn't convince *me*," Jesus said.

"Repeat the testing by redrawing right triangles," continued Roger. "You will see that every time you draw a right triangle, no matter how squat or tall, the theorem holds. You should become more accepting, but not entirely persuaded. What you need is some more 'evidence,' not a formal proof in the mathematical sense, but a 'somewhat credible' argument." He sketched several triangles and said, "Examine these triangles and you will become persuaded that the theorem is true."

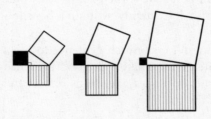

"I'm not persuaded," I said. "Suppose the triangle has two equal sides. Does it work for such a triangle?" I drew a picture.

Form two squares on the sides of the isosceles right triangle and one square on its hypotenuse.

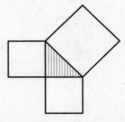

Divide the smaller squares in half to make four isosceles right triangles.

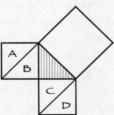

Fit the triangles into the larger square to fill it exactly.

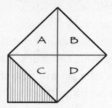

"Okay, I'm persuaded that it works for the isosceles right triangle," I said. "Now, what others can we be sure of?" I drew a triangle and played with it, but I could not get anywhere.

Frustrated, I began to doodle.

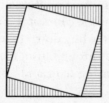
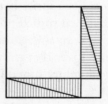

Jesus took one look at my doodling and shouted, "You proved it! All you have to do is rearrange the four triangles in the big square to make the two smaller squares!" And he made the following sketch.

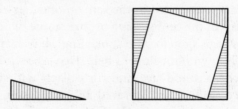

The four shaded triangles can be repositioned from one figure to the other to show that the two squares in the figure on the right must equal the square C in the figure on the left.

It was brilliant. My sketch was nothing more than doodling, yet Jesus saw something I did not. This was by no means a formal proof through a string of logical arguments starting from axioms; nevertheless, it seemed to me to be an airtight argument.

The next day, Jesus told us that he had had a dream that the theorem was not specifically about squares. And when he awoke, he realized that his dream was right.

"It seems that the Pythagorean theorem is about numbers—that is, the squares of the lengths of the sides," Jesus said. "But the square of a length is also the area of a square whose side is that length. So, is the theorem about numbers or geometry?"

"Perhaps it's about both," I said, amazed by the thought and innocent of the remarkable possibility that geometry and arithmetic could reflect each other.

A few days into our trip, we came to the little town of San Mauricio. What a great surprise. The sidewalks on both sides of its short main street were paved in the same mosaic tesseral pattern as those of Caracas. In the center was a bronze statue of Simón Bolivar. The mayor was Jesus's uncle, who spoke English well enough for me to understand him without Roger's help. He was a short man with a bushy moustache, a round face, and curly black hair who constantly chewed tobacco and never removed his sunglasses, even as the evening turned dark. He arranged a feast in our honor. Most of the town was visible from our outdoor feast on a little hill under the shade of tall mahogany trees, with nearby monkeys playfully jumping in all directions. In the distance, I could see long, low buildings that might have been army barracks; three gray water towers; and two rows of thatched roofed mud houses, *churuatas*. I could see the town communal vegetable garden sprawling down to a wide section of the Manapire River that zigzagged through lush, receding valleys with birds of all colors tunefully sharing fruit and berries of their heavenly habitat.

A magnificently embroidered tablecloth of geometric design was laid out on three long boards supported by tree stumps. Exotic fruits of every kind were placed around the ends to keep the cloth from the welcome evening breezes. Apricots, mangos, and guava were piled high in the center, along with several cooked chickens bathing in thick coconut milk. Two large goldfish bowls of lemonade and cut lemons balanced the boards. My thirst was perpetual, and although I recall the lemonade more vividly than anything else on the table, I still remember the flowers and rainbow-colored berries strategically placed to make an impressive presentation. It was quite a surprise that such a remote village would salute unimportant, uninvited, undeserving strangers with such generous display.

Our host was very interested in America, especially President Kennedy. Conversation was a battle. Although he seemed to be an educated man, he had the charmingly innocent background of someone with only Caracas newspaper and radio versions of the bigger world outside of his remote village.

Jesus, meanwhile, couldn't get enough mathematics. He returned to his thoughts about the Pythagorean theorem and felt that somehow he had to be more convinced. I had no real idea how to prove the theorem in any conventional sense and could only try to persuade him by virtue of experience, intuition, and compelling forces suggesting that it couldn't not be true. Parmenides was referring to this kind of persuasion in his poem *The Way of Truth*: "The only ways of enquiry that can be thought of: the one way, that it is and cannot not-be, is the path of Persuasion, for it attends upon Truth."

I was amazed to find that even small schoolboys in San Mauricio knew the Pythagorean theorem and could give reasonably persuasive arguments in favor of belief. The mayor called for a boy, no older than ten, to give credible evidence that the theorem is true and to show how well his town's school prepared schoolchildren. The mayor pretended to be choosing a boy at random from a small group of curious children standing at a comfortable distance behind an imaginary dividing line. The boy was dark-skinned, shirtless, and smiling, and had hair combed in an attempt to cover a recent wound to the left side of his head. With a great deal of help from the mayor, the boy allegedly proved the theorem while drawing a picture on the back of an old movie poster of Clark Gable hovering over the exposed cleavage of Vivien Leigh.

The schoolboy had given a well-known Hindu "proof" of the Pythagorean theorem. The Hindu version simply gives a picture like

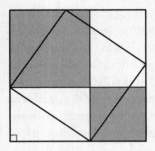

the one the boy drew, except that it includes the word *BEHOLD!* It is a very persuasive picture. From it, you can deduce the Pythagorean theorem by comparing the areas of the triangles, rectangles, and squares in the figure.

The problem with this "Hindu proof" is that it is not really a proof, but merely a picture. The proof is in the ability to rearrange the triangles, rectangles, and squares in such a way that the two small (shaded) squares fit inside the tilted square.

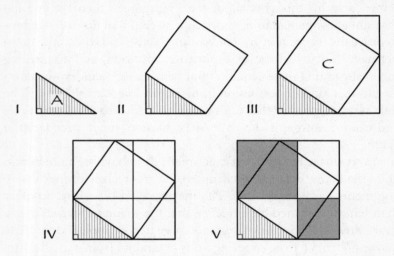

This is essentially what the boy said: "Draw a right triangle. (I). Then draw a square on the hypotenuse (II). You can fit both your triangle and the square on the hypotenuse into a larger square (III). From two corners of the original triangle, draw two lines parallel to the sides (IV), and color or shade the four rectangles defined by the large square and the two lines that you just drew. Your picture should look something like V."

In Figure V, there are four triangles that have the same area as the original triangle. There is a tilted square, which is really the square on the hypotenuse of the original triangle. And there is a big square that borders the entire figure. Next, notice that there are two small squares that are the squares on each of the sides of the original triangle. Let A represent the area of the original triangle. Let B represent the area of the big square.

Check that $B - 4A$ equals the area of the tilted square. Then check that $B - 4A$ also equals the sum of the areas of the squares on the sides of the original triangle. Now use the fact that two things that are equal to a third must be equal to each other to convince yourself that the theorem is true for your original triangle.

The next morning, the mayor, encouraged by the schoolboy's performance, took us to the village school that was established in 1902 and appropriately named, in translation, The Search for Knowledge School. It was impressively active; students were seated at desks working on what appeared to be math projects. The teacher seemed to know the history of mathematics better than mathematics itself, but he was clearly attached to his students. He was a tall man, whiter than his fellow villagers, with a small black beard and a nervous habit of using one hand to pick at the fingernails of the other. He claimed to have spent some time in Texas doing migrant farm work. His English was clear. When I asked about textbooks, he exchanged glances with the mayor and said, "I am the only one with a textbook. We don't have enough textbooks; so I give the lessons from my copy." But the mayor interrupted.

"We have a fine library," he said and immediately led us out of the schoolhouse to proudly lead us to one more place.

How to Persuade Jesus

Is the Pythagorean Theorem True?

"As the scale of the balance must necessarily be depressed when weights are put in it, so the mind must yield to clear demonstrations." The more the mind is empty and without counterpoise, the more easily it yields under the weight of the first persuasion.

—Montaigne quoting Cicero, Essays

The mayor was right to be proud. The town library was a one-room wooden bungalow, painted on the inside from floor to very low ceiling in lively colors. Two windows and the front door welcomed the daylight that reflected off a mural mimicking the real windows and door. It was painted with such skill that the brightness of the painted light suggested that we were looking outside at the huge ferns that really were just on the other side of the wall. The books were old and in bad physical shape, leaning at various angles against milk crates on low bookcases. Two metal wire chairs with round bottoms faced the windows. They were the uncomfortable kind, popular in American ice cream parlors at the time. A five-pointed star inscribed in a circle was faintly painted on the floor and could barely be seen under heaps of scuff markings.

Humidity and mildew had stained many books. Others were in worse shape, having been unfortunately stored more directly under leaks in the roof. But pick up any book on any shelf, and it would hold your interest until you began to think that you should be browsing rather than reading. I spotted an 1889 math book written in English. Its title summoned me—something containing the name Euclid. The musty book, with its curled hard cover, cracked open to a page containing the same diagram that the schoolboy drew on the back of the *Gone with the Wind* poster. A neatly folded note fell from the book as I opened it to where it wanted to open: to an illustration almost identical to the one I doodled, the one Jesus claimed to be a proof. It seemed that someone had tried to work out the proof from the illustration.

I borrowed the book to study it further. Jesus did not know that the Egyptians had discovered his proof long ago. The book claimed that there was little doubt that the Egyptians were dissecting figures that enclosed areas and playing with general ideas in geometry long before Greek travelers came to the North African continent.

The mayor left us back at the table under the mahogany trees, where I rested and read my borrowed book while the villagers went about and a few preschool children watched from a distance. I was happy to have one more book written in English. I was surprised to learn that the Egyptians had very persuasive arguments supporting general theorems, although I felt sure that such arguments fell far short of Greek standards of rigor. Unlike the Greeks, who went by cult names such as Pythagoreans and personal names such as Thales, these Egyptians remained nameless. Histories refer to them as simply "the Egyptians" and give no time period for their contributions. They could have made contributions at any time from the building of the Great Pyramid of Gizeh to the end of the Trojan War. Quite possibly this is because in Egypt and more ancient civilizations, special knowledge of writing and mathematics was limited to the priesthood, which monopolized and preserved learning as collected sacred dogma.

The book told tales of Thales as if the author personally had known him. It painted a picture of a man sailing back and forth from his hometown of Miletus to Egypt. He was in the business of buying and selling papyrus, farming tools, boats, olive oil, and olive presses.

While in Egypt, he met with a priest who took him to see the pyramids at an inspiring moment when the sun cast a sharp shadow along the golden desert floor. At that moment, Thales had an idea. Could he determine the height of the pyramid just by measuring shadows? He poked a rod into the sand and waited for the moment in the day when the length of the rod's shadow cast by the sun would also be equal to the rod's height. That would be the moment when the length of the pyramid's shadow would also equal the pyramid's height. He was quick to turn particular observations into general truths.

Aristotle tells a tale to demonstrate Thales's shrewd business sense. Eating fresh figs under a barren tree in an olive grove, Thales realized that a tree's abundance of fruit depended on the weather and that weather follows cycles. He then thought about how to use astronomical observations to predict favorable weather conditions for olives and bought up almost all the olive presses in Miletus to corner the market for substantial financial gain. We learn charming anecdotes about Thales from Proclus, Plutarch, Aesop, Plato, and many others, but we still know very little about the man himself. Aesop tells a tale of Thales's donkey: By accidentally rolling over one day in a stream, it learned how to lighten its load of sacks of salt. Thales's clever solution was to load the donkey's sacks with sponges. The stereotype of the absent-minded professor may be attributed to Plato's tale of Thales walking into a ditch. One of his attendants called out, "You know what's happening in the sky but can't see what's happening at your feet."

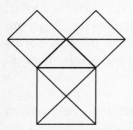

When Amasis, King of Egypt, heard about Thales's clever technique for measuring the heights of pyramids, he invited him to the palace. An artist was working on a mural in one of the great halls. He had prepared the wall with square markings to line up and orient his figures. Those markings inspired Thales's next idea. He noticed that he could prove a special case of the Pythagorean theorem (in which the right triangle has two equal sides) by simply drawing a picture similar to the palace artist's markings. (This is the same as the proof described on page 16.)

The Egyptians knew that triangles with sides of specific meas-urements—such as three, four, and five—would have to be right tri-angles, but it's doubtful that the Egyptians knew the full general force of the theorem before Thales's time. Proclus, the philosopher-historian who wrote almost a thousand years after Thales, attributes several important mathematical facts to Thales. He writes that an ox was sacrificed after Thales discovered that an angle inscribed in a semicircle is a right angle. However, such stories of sacrifice are also attributed to the Pythagoreans on their discovery of their famous theorem, contrary to their cult's laws prohibiting them from eating meat. Proclus also tells us that Thales discovered that a circle is bisected by its diameter, but we don't know how he proved it.[1]

The picture in the book that I was looking at was the one Thales is reputed to have drawn and was very similar to the one I had drawn when I had shown that the Pythagorean theorem was true for the isosceles right triangle.

Whose proof was it, Jesus's or mine? I had simply doodled, and Jesus, inspired by my scribble, saw a proof. It was Jesus's because he used the drawing to make an argument (the rearrangement of the triangles) that showed the Pythagorean theorem to be true. His proof was reminiscent of the Hindu proof the San Mauricio school-boy presented with his drawing on the back of the *Gone with the Wind* poster, but more geometric. The Hindu proof requires a bit of arithmetic—one must subtract and compare areas to be convinced that the theorem is true.

My peaceful reading ended abruptly when the village school let out; a mob of kids encircled me, giggling. I escaped to find Roger out on the narrow streets at the edge of town where the dusty roads faded into stony tracks that narrowed to grassy paths through forests.

Late that morning, we left San Mauricio. Roger and I held vertical poles in our army truck as it careened from side to side on the muddy road to military camp in Cabruta and the Orinoco.[2] Miguel ordered Roger and me off the truck before we came to the town, fearing that his commanding officer might see us. We walked the last several miles to town. A lost capuchin monkey, resembling a monk dressed in a cowl, playfully scuttled alongside us while I confessed to Roger my apprehension. Roger spotted toucans and macaws in dragon trees while I talked. He cracked his knuckles, in that same annoying way he was wont to do, and launched into one of his customary soliloquies; this time he unleashed the story of Richard Burton and John Speke setting off from Zanzibar in 1856 for an adventurous expedition to find the source of the Nile, where "no white man had ever been before."

We sat on the side of the road, as we often did, writing in our journals and eating dried fruit as a cute little kinkajou played in a nearby tree. I wrote sketchy, fragmented sentences in my little notebook to log my observations; Roger wrote poetry.

He would whip out lines of splendid poetry without making changes, as if his words flowed from a mind trance of images before a screen and he was merely a scribe. His words came as fast as he could write. And when he finished a page, he would tear it from his notebook and throw it away, even as he continued the same poem. I would pick up his papers, sometimes using my feet to catch them from the wind, and save them in my knapsack.

After consulting our maps and continuing our walk to town, he conceded that it would be too difficult to get to La Esmeralda. But that didn't stop him from continuing the story of Burton and Speke.

"Do you see that?" he asked, interrupting himself. He was pointing in the direction of what looked like a maize field.

From a distance, we could see a group of naked children mimicking playful spider monkeys in a dragon tree and a drove of grazing guanaco nearby. The children were *Panares*, the indigenous peoples of the region. A little farther along we saw a group of older Panares women with one young one, who was beautiful and bare breasted. She was carrying several very colorful baskets of geometric design. Roger looked at me, cracked his knuckles, and asked me if what we were looking at was real. He approached the women, who giggled and backed away. But as the children ran to join them, they gained enough courage to permit Roger to approach and indicate that he would like to buy a basket. He gave the young Panares a few

Bolivar too many; she returned his kindness with a carafe of *cachiri*, a drink made from sugar cane and fermented maize.

The sun was still high in the sky, and the heat was scorching when we reached the outskirts of Cabruta and its narrow dusty streets. Our water supply was gone, and the carafe of cachiri almost empty. The tranquility in the streets leading to the main square contributed to our eerie feeling of being intruders, trespassers, from another world. It was siesta time. We passed the curious stares of a few old men sitting on wooden chairs in front of their churuatas as we nervously walked to the center. The Church of San José, surrounded by shade trees, faced a square in the center of town. It was a simple tan adobe building, with an ornate but crumbling bell tower that had seen better days. Inside, natural sunlight fell through openings in cupolas onto painted images of the stations of the cross against plain white walls, inviting auras of meditation.

We sat on wooden pews, resting and contemplating the pain and sacrifice depicted in the stations, as if we were invited to be there. Then Jesus threw open the front door, excitedly running to tell us that he had a simple, airtight proof of the Pythagorean theorem.

"Jesus," I said, "let's eat and drink first."

Jesus had relatives in Cabruta. Another uncle, Jorge, lived in an intricate tin-roof house at the edge of town. His aunt, a thin, smiling woman with dark blue eyes, prepared a meal of bass and potato soup with cornmeal bread while Jesus showed his proof. Jesus put down a small pack of curled paper on a table just outside the house, as a growing number of curious town children and assorted pets gathered to inspect the two pale foreigners.

"Watch this," he said with a smile while drawing on one of the papers. "It isn't about squares on a right triangle. It's about proportions. Take a blob like this," he said drawing a right triangle *ABC* and a blob next to it. "Draw side *AB* of the triangle as a line somewhere on the blob.

Now scale up the blob so the line *AB* becomes the same size as the base *BC* of the triangle." He drew the new blob.

"Scale the blob again so that *BC* becomes the same size as the hypotenuse *AC*." He drew a third blob.

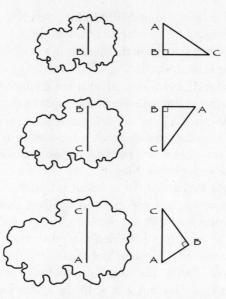

"Okay! The area of the first blob plus the area of the second blob equals the area of the third," he said with an open-mouthed smile that showed two gold teeth.

At this, Roger managed to continue talking as if he were translating. But Jesus was not saying anything.

"Take any right triangle," he said while drawing a right triangle with a single line from the right angle to the hypotenuse.

"There," he said conclusively. "That'sh the proof. No one can doubt that the shum of the triangles on the shides is equal to the triangle on the hypotenushe because the two shmaller triangles on the shides are shitting inside the large triangle perfectly!"

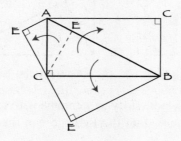

Flip the three right triangles, along their respective hypotenuses, from inside triangle *ABC* to outside triangle *ABC*.[3]

The lisp was not the problem. I just didn't understand the math, though it seemed that Jesus did.

"Yes! Of course!" Jesus said. "The small triangle is just the first blob, the larger one is the scaled-up blob, and the original triangle is the biggest blob!"[4]

"But still," I said, thinking there was something missing, "you proved that if your blobs are right triangles, it works. But how do you know it works for other things?"

"Aha!" Roger exclaimed. "It is about proportion. Watch thish." And at that, he wrote out the math to confirm that his argument proved the classical Pythagorean theorem for squares.

That was the first time I learned that most elegant, slick proof. Years later, I learned that Einstein had discovered this proof when he was a young man. But either Roger discovered it independently of Einstein or he recalled it from something he saw before.[5] The ancient proof of the Pythagorean theorem is contained in Book I of *Euclid's Elements*. It is the forty-seventh and next-to-last proposition of the book; it takes twenty-eight of the previous propositions to prove it. Indeed, the proof is solid and elegant. But do we really need twenty-eight propositions to be convinced? The proof that Jesus and Roger showed me was elegant and overwhelmingly convincing.[6]

If the Hindu proof is in the picture on the left, then the Western proof is in the picture on the right.

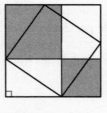
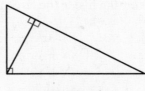

BEHOLD!   BEHOLD!

Behold tells the whole story to someone with enough mathematical experience to know that the Pythagorean theorem is not about squares, but rather about figures whose areas relate to each other as the sides of the triangle relate to each other. In other words, if you have proven the theorem for one geometric type of figure (say, triangles), then you have proven it for any other geometric type (say squares, circles or blobs).

One might wonder why Euclid constructed a proof so complicated when Einstein's proof seems to be thoroughly convincing. Consider Einstein's proof once again. When we extend the proof from triangles to other shapes—squares, in particular—we find that we need to first know something about proportions and then know how to use algebra to manipulate ratios to see them translate into squares. We innocently—but rightfully—neglect to reflect on what gives us permission to manipulate symbols in the way we do. In the end, facts about proportions plus the validity of the kind of symbol manipulation used in Einstein's proof might require as many propositions as Euclid's forty-seventh proposition.

Soup and bread were ready. Most household functions, such as cooking and sewing, were done in a courtyard flower garden just beyond the front door. It seemed odd that the garden plants were identical to the flora of the town's surrounding habitat, but there they were, miniature palm trees, arboreal ferns, and other bromeliaceous jungle plants with spiny leaves, dense spikes, and multicolored bracts growing in clay pots of all shapes and sizes. In contrast to the smells of orchids commonly found in the surrounding forests, they supplied a wealth of scents to compete with the cooking of onions and basil in preparation for the fish and potato soup we were about to eat.

Roger told of our plans for adventuring up the Orinoco to a garden full of blank stares, unsure expressions, and short laughter. There was one old man standing very close, smoking a long pipe with smoke so powerful it irritated my eyes, even though we were outdoors. "You can't head up the Orinoco!" he said. "You need food, guns, and a guide! How do you expect to make your way through the jungle? There are no roads, and you will get lost without a guide. And who knows if there are any guides who know the south country?

There are snakes, jaguars, pumas, ocelots, and unfriendly natives. If the animals don't tear you apart, the snakes will poison you. And who knows what the Yanomami will do?"

These words of wisdom discouraged Roger. He cracked his knuckles.

"Why not head downstream instead?" said an ageless bearded man chewing tobacco while standing and stroking a large hound near the garden entrance. "There are many spectacular waterfalls in that area, like the Uraima on the Paragua River, or La Hacha and El Sapo."

"And less danger from wild land animals and alligators," said Jesus's uncle, Jorge.

"There is a legend," said the bearded man, "that an American landed a plane high in the mountains near Auyan-tepui to look for gold and diamonds, but he couldn't get the plane off the mountain and it's still there. He discovered a great waterfall."

"Yes, yes," said the other man, "a mile high. His name was Angel; the falls is called 'Salto del Angel.'[7] There are many spectacular falls in that area."

Roger turned to me. "Auyan-tepui means devil's house," he said with gleaming eyes. He pointed to the Auyan-tepui on his map, circled it with his pencil, and then turned to the bearded man and asked, "How can we get to Auyan-tepui?"

Someone in a remote corner started strumming a two-string mandolin. For a few moments, all that could be heard was its sound. After a short overture, the entire party began to hum. They sang songs, ate fish and potato soup, and drank a strong homemade wine tasting much like cachiri until the hot sun finally set in a purple sky. The air cooled, and the party continued by the light of several kerosene lamps and what must have been a hundred thick candles.

Early the next morning, I woke to the sound of another mandolin and a woman humming a song about coffee. The strong smell of coffee was everywhere in Cabruta, and it was market day, the day when everyone came to the village square to loiter, exchange goods, and drink coffee bought from competing coffee roasters. By midmorning, coffee smells faded into a mixture of lemon and thyme or dried fish, depending on the shifting breezes. One could buy matches, cheap jewelry, sunglasses, and colorful fabrics, as well as mangoes, bananas, tobacco, and sugar cane. One stall sold parrots, monkeys, and baby ocelots; another sold river dolphin. Roger and I drank cup after cup of coffee in the market square, shooing flies while planning our next adventure.

"Look at this fountain," Roger said, pointing to the fountain in the middle of the square. "It's a pentagon, not a hexagon."

"So what?" I responded.

"Fountains are usually hexagons," he explained. "Hexagons are easy to construct. You just take six equilateral triangles and fit them together. But a pentagon…."

He was interrupted by the sound of a motorbike racing up and down the narrow road leading to the square.

"Do you know the story of the Pythagoreans, the brotherhood of followers of Pythagoras?" Roger continued. "A covenant bound them to secrecy over their master's teachings and anything else taught or discovered by the brotherhood. One of their secrets was the construction of the regular pentagram, the five-pointed star and symbol of the brotherhood. It's not easy to construct if the only tools permitted are a straight edge and compass; in other words, can you construct a regular pentagon[8] using only straight lines and circles?"

At that, Jesus found a few more sheets of newsprint-quality scrap paper of assorted sizes and started sketching. So did I, but I quickly gave up.

"Roger," I asked, "do you know how to do it?"

"Yes," he answered. And without hesitation, he began to explain his method. "You only need to construct an isosceles triangle with one angle equal to half the other two. If you do, you will have constructed an angle of 72°, and that is exactly what you need to complete the pentagon because a pentagon has five sides, and the sum of all the angles of a pentagon is 540°. Euclid showed us how to do it."[9]

It took a long time to digest Roger's construction; I had to ask a million questions and have him repeat himself at every stage. But after working it out myself, it became so clear, so right, so evident that I was persuaded that I had known it all my life.

Roger, satisfied with the success of his lesson, decided to tell yet another mathematics story. It was about a river dividing two countries. The river followed such a complicated labyrinth that the explorer had a hard time keeping track of which country he was in after he crossed the river many times.

"If you were to pick an arbitrary point on the map, it would be hard to tell what side of the river it is on," Roger said. And with that he told a story of Camille Jordan, the nineteenth century professor of the Collège de France best remembered for his proof that a curve that doesn't cross itself and begins and ends at the same point divides a plane into exactly two regions.[10] As with his account of the discovery of Fermat's Last Theorem, Roger—a poet at heart—took some creative license with the incidental details of the story.

CHAPTER 3

The Simple and Obvious Truth

The Role of Intuition and Belief in Mathematics

"Just the place for a Snark!" the Bellman cried,
As he landed his crew with care;
Supporting each man on the top of the tide
By a finger entwined in his hair.

"Just the place for a Snark! I have said it twice:
That alone should encourage the crew.
Just the place for a Snark! I have said it thrice:
What I tell you three times is true."

—*Lewis Carroll*, The Hunting of the Snark

On a drizzly October afternoon in 1886, Camille Jordan entered a small building behind the Pantheon in Paris to deliver a lecture to his mathematics class at the École Polytechnique.[1] He had just returned from a circuitous walk to and from the Café Luxembourg several blocks away to retrieve an umbrella that he had left behind. Such short walks to or from his favorite café often inspired clever ideas to help streamline his work or lectures; this time, his idea was magnificent. He entered a small mahogany-paneled classroom through a door reserved for professors and faced thirty-one students seated in a steep incline of rows. He had taught his famous class many times before, but on that particular afternoon, encouraged by his latest idea, he confidently said something that had an unexpected consequence. He intended to prove a theorem by means of a statement that he had always thought to be obviously true, so he casually relayed it to the class. A vigilant student, seated in the last row, politely interrupted the great professor to ask for more evidence or a proof of what was claimed to be "obvious." Professor Jordan scratched his head, stroked his beard, and rapidly blinked his eyes as he nervously removed his wire-rimmed glasses from one ear at a time and thought about how he would convince the class that the simple statement he had made was, indeed, true. After pondering the statement more carefully for several minutes without saying a word, he concluded that perhaps it was not so obvious.

As the hour progressed, the professor's simple statement fell from the obvious into an abyss of vexing uncertainty. The course turned from its usual syllabus to the professor's not-so-obviously-true statement, which turned more and more elusive at every class meeting. Jordan spent the year routinely working each day at the Café Luxembourg, consuming coffee and a large number of croissants on which he spread spoonfuls of peach jam. Proof eluded him. In the spring, he explored new walking routes. From the École Polytechnique, he would walk eastward along Rue des Boulangers, smelling the breads that had continuously been baked and sold throughout each day since Napoleon returned in defeat from Russia. He discovered the Jardin du Roi by walking farther eastward to the Riviere des Gobelins, a small tributary to the Seine. In the garden, he walked amid the scents of rosemary, thyme, and medicinal plants still growing from the days when the park belonged to a school of botany. Inspired by the beauty of nature and a feeling of being in the country, he would sit on one particular bench by the river to watch swaying laurel leaves while thinking through his proof. By the following June, he had completed the proof and written it in several hundred pages of notes that were published eleven years later in a 109-page compendium to a three-volume text of his lectures. The simple and obvious statement that Professor Jordan casually made in October 1886 was, indeed, true—at least, for the time being.

The statement was the point of the story Roger Hooper told the day before he set off down the Orinoco. I didn't see it so well at the time, but he was deeply concerned with belief, persuasion, and proof. That was evident that day on our long walk to Cabruta after encountering the Panares women and children, when he questioned the truth of the images he had seen before his own eyes. We were tired, hungry, thirsty, and uncomfortable in the humid heat of the rainforest. Whenever Roger was uncomfortable, he felt a need to talk, and puzzled by what he had just seen, he speculated with great seriousness on the topic of belief and persuasion. Much of what I have to say comes from him.

We often hold opinions without knowing why, and presume them to be true without having definite proof. But proper proof is a process that can either change an opinion or stiffen it into unyielding persuasion. At some unnoticeable point in that process, we start to "feel" the truth. While a mathematician learns a proof of a theorem, subconscious links are slowly formed between what is being proved and an intricate, gigantic web of connections to old, established theorems in his or her cache. Professor Jordan's feeling of truth for his statement came long before he had any argument to back it up. How could he have known that the statement was true nine months before he was able to make any conclusively deductive arguments? Mathematics enjoys the honorable distinction of being free from judgment; yet even mathematicians often form strong opinions without bothering to link them to the usual deductive network of known mathematics. Camille Jordan struggled with the forces that separate unsubstantiated opinion from proven fact. Driven by his unconscious instinct to substantiate his hunch, he had to find a way to prove he was right; he started by confusing his initial opinion with a "feeling" of truth and tried to communicate that feeling to others.

The simple statement that Camille Jordan had innocently thought was so obvious when he first used it in his class is the following: "Every continuous curve that begins and ends at the same point without crossing itself divides the plane in two." Surely, a circle is a specific example of the kind of curve Jordan had in mind; it fits the description as a curve that separates the plane into two regions, those points that are inside the circle and those that are outside. If we think about it for a few minutes and perhaps doodle with pencil and paper, we are likely to agree that it is obvious that every continuous curve that begins and ends at the same point without crossing itself divides the plane in two.[2] But mathematicians are very suspicious of claims of the "obvious" without some hard evidence of "truth," even though, more often than not, their intuition turns out to be right. Mathematicians seem to sense mathematical truth. They might use the word *obvious* to communicate a strong belief that a formal proof can be found lurking behind a heuristic argument; after all, the whole notion of proof in mathematics has never been clearly defined. It is normal for mathematicians—and everyone else—to make true statements long before formal proofs are found. But sometimes the omission of a logical detail backfires.

Jordan initially thought his statement was obvious. What could that mean? I suppose he thought it required no thought or consideration for the mind to accept it. Perhaps he initially thought it hard not to easily sense its truth. To him, its truth was clear and apparent, as if he could sense it with his own eyes. But even truths that are seen through the eyes can be called into question. When Galileo discovered four new moons orbiting Jupiter, he was admonished because he had observed them with the help of a telescope and had not deduced them from logical arguments. Here is a case in which someone is seeing the moons of Jupiter and is told that what he is seeing cannot be true because logical argument is better than direct observation. The Florentine astronomer Francesco Sizzi mocked Galileo:

> The Jews and other ancient nations, as well as modern Europeans, have adopted the division of the week into seven days, and have named them from the seven planets: now if we increase the number of planets, this whole system falls to the ground…. Moreover, the satellites are invisible to the naked eye and therefore can have no influence on the Earth and therefore would be useless and therefore do not exist.[3]

Galileo was actually seeing the moons of Jupiter with his own eyes, albeit through a telescope. Anyone could have seen the moons, had he or she just looked. Almost four hundred years ago—not so long ago—persuasion by logical argument or philosophical principles was considered stronger than persuasion by direct observation. Even seeing through the human lens and retina would have been considered indirect observation, so, surely, looking at the moons of Jupiter through a telescope was even more indirect. Yet, for us, we accept what we see as if it were truly direct observation. What do we see through night-vision goggles, now used by the military? Soldiers depend on the images they see; lives depend on them. And yet, all they are really seeing is infrared radiation—heat, not light. We even go so far as to trust radio waves to bring us pictures of the moon, of Mars, and of the bottom of the Black Sea. Why aren't we more doubting?[4]

When Roger Hooper saw that bare-breasted young Panares woman in the diorama foreground of dragon trees, guanaco, and naked children mimicking spider monkeys, he asked me whether she was real, as if he should question what his eyes told him. It is a natural question for a mathematician in the forest because we can get so involved in seeing ideas through our minds' eyes that we overlook the natural world. Yet, after all due consideration, we confirm or deny facts according to what we see or what our other senses tell us.

Belief can alter memory, and, conversely, memory can drive belief. Belief itself is not only strong enough to change memory, but powerful enough to influence future successes. Leonard Shelby, the protagonist in the film *Memento,* shares his thoughts about this: "Look, memory can change the shape of a room. It can change the color of a car. And memories can be distorted. They are just an interpretation, not a record; and they are irrelevant, if you have the facts."[5]

Many years ago, I wrote a short story about a man, Mr. Unis, who performed an extraordinary stunt on a street corner near where I once lived. He worked for the circus and had the astounding ability of being able to stand on one finger. As I was writing, I believed that I had truly encountered him in just the way I was writing it. But now as I think of it, I am not at all sure that I really did encounter the man. Could my memory have played a trick? Could anyone possibly stand on a single finger? Was it a dream? Could there have been a Mr. Unis?

My memory seemed to have concocted a quasi-mythical figure.

Several years after writing my story, I attended a circus perform-ance: It was a special performance on the hundredth anniversary of the Ringling Brothers Circus. I was seated high in the stands of the Boston Garden (now the Fleet Center), where, once again, I saw the very same Mr. Unis. He was not only standing on one finger, but also balancing himself on the end of a thirty-foot pole. I was astonished. My quasi-mythical figure was not a fantasy! Although I still found it difficult to believe that my meeting with Mr. Unis had been real, I became more convinced that it had been. Still, there was a shadow of a doubt caused by the presumed physical impossibility of the stunt. When I was younger, I could have believed that a man could balance himself on one finger. Yet once a better knowledge of physics replaced my naïve understanding, it became very difficult to believe in things such as magic, spirits, and humans standing on fingers.

It might seem that we become convinced by repeated exposure to what we see and accept. But it's not that simple. I spent five years taking care of my wife after she had been in an automobile accident with a truck that ran into her. I saw the accident happen and replayed it in my mind thousands of times over a five-year period. What I saw in my mind was an enormous covered truck that was loaded with large boxes. The case went to trial. But in preparation for trial, I was given a picture of the truck. It was a shock to see that it was merely an unloaded 3/4-ton pickup. The truck had such a devas-tating effect on my life that I had "monsterized" its image. It was a fiction of my imagination

The English philosopher David Hume put it succinctly this way: "*Belief* is something felt by the mind, which distinguishes the ideas of the judgment from the fictions of the imagination."[6] "Felt?" What is the sensation of belief? We don't taste it, smell it, see it, or hear it. We don't feel it in the same way we feel warmth, chill, or a toothache. And yet…. And yet we can feel anger, sadness, or happiness. We can feel emotions such as love, hate, and anger. We can even feel impres-sions of approval or insult. But can we feel conviction?

Let me jump now to a time when I first started teaching and experimenting with different presentation styles to demonstrate the pressure of authority. It was early fall in the 1980s. Each morning, I planned my classes while taking a half-hour walk along a well-beaten forest trail from my house to my office, thinking and paying peripheral attention to the surrounding nature. There were hills to climb and streams to cross, with the occasional distraction of a scurrying rabbit or squirrel. Those wonderful walks always seemed to contribute rich ideas for my next classroom presentation. When I had time, I sat on a log to hear the silence of the forest: a thrush foraging in dry leaves, a brief gust of wind rustling weary branches with the last leaves of autumn delicately holding on. One of those walks gave me the idea to perilously push the limits of my professorial license: to intentionally tell my class something utterly false and relentlessly continue to use the commanding influence of my position to push my students to believe it. It was a math class for liberal arts students, an eclectic selection of topics accessible to nonmath majors, a survey of number theory, logic, and several different kinds of geometry. I told the class that the number $2^p - 1$ is a prime number whenever p is a prime number. I then wrote out the calculations for each p from 2 to 19:

$$2^2 - 1 = 3$$
$$2^3 - 1 = 7$$
$$2^5 - 1 = 31$$
$$2^7 - 1 = 127$$
$$2^{11} - 1 = 2,047$$
$$2^{13} - 1 = 8,191$$
$$2^{17} - 1 = 131,071$$
$$2^{19} - 1 = 524,287$$

I had no idea what the response would be. The first three calculations clearly turned out to be prime numbers, but by the time we hit a number in the thousands, students preferred to rely on my authority. I even volunteered to show that 8,191 is prime by using a calculator to divide by all the primes from 3 to 83.[7] This was a magician's slight of hand: No one noticed my intentional skip over 2,047.

As expected, no one doubted that $2^p - 1$ is always prime. Of course, it isn't: $2,047 = 23 \times 89$. But when the dust settled, one skeptic began to have some doubt. Thomas, a tall, thin young man, came barefoot to class each day. He often sat in the first row, sometimes stretching his long feet straight in front of him and sometimes bending his knees to hide his bare feet under his chair. The position of his feet subconsciously signaled his persuasion. He was always alert. On the morning of my experiment, his outstretched feet began to shift into different positions—crossing each other, first right over left then left over right. When I had completed checking that 8,191 was prime, Thomas's knees slowly bent to pull his feet under his chair.

"Thomas," I said, trying to level the playing field of authority. "You seem troubled by what I'm saying. Is there something wrong?"

"I guess...I guess," he shyly answered. "I guess I'm inclined to believe that $2^p - 1$ is prime, but...I'm not convinced without seeing a proof."

I was caught. At that point, I had to concede that I had played a game. It was not true that $2^p - 1$ is prime for all primes p. Indeed, it was not true for $p = 11$ or $p = 23$.

$$2^{11} - 1 = 2,047 = 23 \times 89$$
$$2^{23} - 1 = 8,388,607 = 47 \times 178,481$$

To find the factors of 8,388,607, one needs to divide by only the eight possibilities, 3, 7, 13, 17, 23, 37, 43, and 47 because one of the factors must end in either a 3 or a 7. (Any number ending in 7 cannot be divisible by 5 or any even number. That leaves 3, 7, and 9. Because 9 is already divisible by 3, we conclude that one of the factors of any number ending in 7 must end in either 3 or 7.)

Mathematics has a reputation for being a pure judgment, totally distinct from fictions of the imagination, but is it really? Even in mathematics, experience ignites the process that drives opinion to persuasion. Camille Jordan's initial thought that a simple closed curve divides the plane into two distinct regions came from his extensive experience with simple closed curves. He must have imagined generic curves dividing the plane and subconsciously accepted the division as obvious. It was a fiction of his imagination. He might have even used the statement many times before without paying much attention to the fact that it needed proof. After all, it was obvious and, therefore, didn't require proof. How fortunate that this obvious fiction of his imagination turned out to be true.

Our beliefs might simply be influenced by our culture or our ignorance. Take, for example, "A four-leaf clover is very hard to find in a field of clover." You might be convinced that it is true because you have heard that a four-leaf clover is a lucky find. But have you ever tried to find a four-leaf clover in a field of clover? If you haven't, how do you know that it is hard to find? Andrew Weil relates a wonderful story to suggest that belief is a strong influence on what we see. He writes:

> Years ago I met a woman who was able to find four-leaf clovers in any clover patch. … When I would look through patches of clover, I could search without success until my vision blurred, and whenever I thought I saw four leaves on one stem, they always turned out to belong to two different clovers. But after meeting this woman and watching her do it, something changed for me. I realized that the key to her success was her belief that in any clover patch there was a four-leaf clover waiting to be found. With that belief, there is a chance of finding it; without it there is none. After meeting her, I began to look again, and soon I started to find four-leaf clovers.[8]

When I was a graduate student, an apocryphal story about an extremely brilliant mathematician in the department circulated around the mathematics community. Over the years, I've heard the same story told of other celebrity mathematicians, so it's pointless to name the person behind this story. One day, this genius, an undergraduate at Princeton in the early 1950s, came late to class. His professor had written a list of ten of the most outstanding unsolved problems of mathematics on a blackboard. (The story is apocryphal because it's hardly likely that ten of the most outstanding unsolved problems would fit on a classroom blackboard.) This genius copied the problems, thinking that they were part of the next homework assignment. At the next class session, bashfully embarrassed, he told his professor that he had solved nine of the ten homework problems but could not get the tenth. The point is that the hero of this story had two significant advantages: One was his extraordinary mathematical talent; the other was that he was not handicapped by any severe sense of difficulty. Belief was so empowering that it encouraged him to drive his arguments over the bumps that were obstructions for others who had tried and failed.

In *Alice's Adventures in Wonderland*, when the White Queen says that she sometimes believed as many as six impossible things before breakfast each day by simply practicing belief, Alice laughs and says, "There's no use trying. One can't believe impossible things."

We do not believe in effects that conflict with established experience, so is there a feeling to being convinced of an argument? What did Hume mean when he said, "Belief is something felt by the mind"? One might say, "I 'feel' that I am right" when faced with a belief. Surely, we don't "feel" belief or doubt through physical sensation. The feeling of pain is physical and can be measured by degree. We can describe and distinguish mild pain from torture. To a lesser degree, we can even describe and distinguish pleasure from ecstasy. So is there a "feeling" of being right or wrong? What do I "feel" when I encounter the impossible?

It's not often that one encounters the impossible. So what would your mind "feel" if one day you were having a picnic in the forest and saw a woman riding a horse? Nothing unusual. But what if this horse and rider could be seen only when they were behind a tree and could not be seen when there was nothing to obstruct the view? Such a scene is depicted in the René Magritte painting *The Blank Signature*. Besides being startled, your response would be to deny what you were seeing with your own eyes. In a sense, you would be feeling just what Francesco Sizzi felt when he mocked Galileo's observation of a new planet.

The mind has a big task: to interpret perceptions of the world so that it can live without contradictions. What we see with our eyes does not necessarily have to be truly what exists; it only has to be consistent with our perception of the world. Why shouldn't we expect our mathematical intuition to work the same way?

The Blank Signature, collection of Mr. and Mrs. Paul Mellon. Image © 2004 Board of Trustees, National Gallery of Art, Washington, 1965, oil on canvas.

Just when Camille Jordan thought he had completed his lengthy proof, Giuseppi Peano, teaching at the University of Turin, invented a peculiar curve. Professor Peano's strange object satisfied all the traditional requirements of the definition of a curve but was so peculiar that even Camille Jordan had not accounted for it in the proof of his theorem. Peano's invention put a hole in Jordan's proof. So Jordan went back to his favorite café and scribbled out another hundred pages of proof to accommodate Peano's invention. Again, what

Jordan thought was true by intuition in October 1886 was true by means of a formal proof—at least for the time being. And once again, the story does not end here.

On first glance, you might say, just as Jordan did, "Of course this theorem is true!" The trouble comes when you try to use your intuitive experience of space to prove something that depends on precise definitions. We tend to confuse the spatial model of a curve with the abstract definition of a curve. We haven't seen all the possible contortions that Jordan's curves could go through. But more important, we don't even know what a curve is.

Professor Jordan thought that his proof was correct. He accommodated Peano's curve in the latest version of his proof and felt convinced that all was well. A year passed while his proof circulated through the mathematical community. Alas, just when Jordan was feeling quite good about his theorem, David Hilbert, chairman of the Mathematics Department at the University of Göttingen, invented another anomalous curve that Jordan had not considered. Again, the proof was destroyed.

The chronicle of Jordan's proof continues with several distinguished mathematicians of the time giving proofs that were eventually found to be incorrect. Oswald Veblen was twenty-five years old when he went to Princeton from Chicago. In 1905, nineteen years after that one student in Professor Jordan's class hesitated to accept the statement as obvious, Veblen gave the first correct proof. It was only fifteen pages long.[9] Once again, the theorem was shown to be true. But hadn't it been true all along? And if it had always been true, how could Jordan have known it to be true without having a solid proof?

Seventy-five years after Camille Jordan finished a proof of his theorem on curves in the Café Luxembourg, I found myself in the same café with another great professor. As an American studying in Paris, I had not been informed of the great reverence that the French have for professors and was dumb enough to simply confront mine with questions. Normally, a professor would lecture to a very large audience after entering the lecture hall from an anteroom at the back of a stage filled with microphones. At the end of the lecture, he would leave through the same anteroom, not questioned and not seen again by the students until the next lecture.

One afternoon, in October 1961, I was walking along the Boulevard St. Michel with several French friends when I spotted Professor Roger Godement walking toward us. My friends were so nervous about the proximity of God himself that they immediately crossed the street to give him more space. On seeing this strange foreign behavior, I decided to play it for what it was worth and stopped Professor Godement in his tracks. Of course, he didn't know me. I simply said what any courageous American student would say upon meeting his or her professor: "Hello, Professor Godement. I am a student in your algebra class and wonder if I could meet with you sometime to clarify the proof that you talked about in today's class."

To my great surprise, his answer was what any reasonable American professor would have given, under the circumstances: "Sure, why don't we have a coffee in that café across the street, and I will tell you all about it?" Well, perhaps reasonable American professors wouldn't say this, but at least in America it would not be a surprise if one did. I was not prepared for such a generous offer, but I immensely enjoyed witnessing the shock and surprise portrayed on my friends' faces as the Great Professor and I crossed the street and entered the Café Luxembourg.

We sat in the café for more than an hour while Professor Godement explained the proof in detail. At that first meeting, he was as nervous as I was, but we met at that same café several times during the year. I believe we mutually enjoyed our meetings because he had a rare chance to talk with an undergraduate, and I had the chance to show off to my friends. One day Professor Godement turned to me and calmly said, "You know, about seventy-five years ago, Camille Jordan proved this very same theorem in this very same café."

The story of Camille Jordan's quest for proof, as told by Roger Hooper, would have been totally forgotten had Roger Godement not revived it on that afternoon in October 1961. I hadn't believed Hooper's version until the first time I ordered a croissant at the Luxembourg: Without my asking for it, the waiter brought me peach jam.

My first serious introduction to modern mathematics was through Professor Godement's course in modern algebra at the University of Paris. He stated theorems and proved them, one after another. For the first six months of attending his classes, I felt a strange sensation of both understanding the proofs and not knowing why. Some proofs were long, others short, but they were almost all informally worded. Occasionally, the words would form neat sequences of statements that would make the proof easier to follow, but mostly they were rough and unpolished. Yet I understood the overall effect of the argument or proof. For years afterward, I puzzled over the question of why I could understand and believe that one statement follows another without being told any of the rules. It seemed as if my brain were wired to accept the rules and that learning them was never necessary. It became clear to me then that there were no fixed rules.

In high school, I was taught to mimic proofs that were spelled out in geometry texts. You know the kind: "Prove that two triangles that have their corresponding angles and sides equal are congruent." I was taught to believe that there is some absolute, structural, logical inference that was necessary to accept an argument as valid. But real mathematicians don't talk in absolute logical structures. Texts in my high school days misled students to believe that mathematical proofs come from a well-defined process, as opposed to an artful means of communication. Mathematicians communicate with one another by vague symbolic gestures that indicate what should follow from what was just said. A squiggly mark might represent an "Abelian variety," which is an object that has a complex definition. Somehow, it immediately makes perfect sense in conversation between two mathematicians who have experience with Abelian varieties, just as a roughly drawn picture of a triangle might represent a triangle.

When I hear some false argument about a triangle, I might not be able to immediately pinpoint what part of the argument is false, but I know that *some* part is false. My brain will not accept the argument if the image I'm being presented with conflicts with the one in my mind. Nobody believes that all triangles are isosceles, yet Lewis Carroll designed a very convincing argument suggesting that they are. (See Appendix 1 for the proof.) His argument is hard to refute, yet when it is given, our opinion is firm: We refuse to believe it. Although the arguments in favor of such impossible statements might be swaying, we are not convinced. Why? Our experience of life is the driving force of our belief systems. We become so accustomed to what we believe that we cannot believe otherwise. This is truer in simple mathematics than it is in other areas of thought; for example, there are many erroneous proofs that $1 = 0$: Some are geometric, some are algebraic, and some are analytic. Outside the world of mathematics, we often form beliefs and opinions that have little to do with factual truth. Scientific truth involves extraordinarily complex procedures and evidence that often surprises even the most experienced investigators.

To refute Lewis Carroll's demonstration that all triangles are isosceles, we need only to exhibit one triangle that is not isosceles, but few persons can pinpoint the place where his argument breaks down. Carroll's argument uses a little trick, well known to magicians. It gets us to look away for a moment. His argument deals with two cases; and it is true that there are only two cases: Either a certain two lines meet or they don't. So we walk away thinking that we have fully investigated the only two possible cases—and we have. But where do those two lines meet? We are fooled into following the picture that *he* presents, with an intersection point *inside* the triangle, but all the while the intersection point might have fallen outside the triangle. In this case, the picture is correct, but it is not the only picture.

CHAPTER 4

What the Tortoise Said to Achilles

Logic and Its Loopholes

To say of what is that it is not, or of what is not that it is, is false, while to say of what is that it is, or of what is not that it is not, is true.

—*Aristotle*, Metaphysics

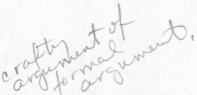

The typical impression of mathematics is that its arguments are syllogistic—that is, they are bundles of "if, then" statements that can be linked together in long chains with a beginning that depends on other valid statements and an end that gives a conclusion. Logic ensures that the links on these chains cannot be broken. But logic can be devoid of meaning, and it alone does not persuade.

We say that an argument is *valid* if the conclusion logically follows from the premises, regardless of whether the premises are true. Examine this argument:

All startbingers are stopbingers.
All gobingers are startbingers.
Therefore, all gobingers are stopbingers.

You would probably conclude that the argument is valid without knowing anything about the truth or falsity of each of the first two statements, even if you don't know the meaning of *startbinger, stop-binger*, and *gobinger*. And consider the following argument:

> All mammals have eight legs.
> All birds are mammals.
> Therefore, all birds have eight legs.

[handwritten marginalia: Sugar is white / Sugar is sweet / Salt is white / Therefore salt is sweet] [handwritten vertical note: law 101]

There is a distinction between the validity of an argument and the truthfulness of any one of its components. Logic is not concerned with the factual truth of the individual statements; it is more concerned with making valid deductions from collections of individual statements. We might extract meaning from the syllogism by replacing it with the following abstract model:

> All *x* are *y*.
> All *z* are *x*.
> Therefore, all *z* are *y*.

Such a model lets one plug one's own meaning in for *x*, *y*, and *z*, and once again shows that the meaning of the individual words are not relevant in examining the validity of the argument. In the end, an argument is valid if and only if the truth of the conclusion is absolutely guaranteed by the truth of the premises.

An *argument* is a group of compound statements. The most interesting arguments are ones containing statements, called *premises*, which logically imply an additional statement, called the *conclusion*. The conclusion is *deduced* from the premises, and the argument is called *deduction*. But the word *deduction* is tricky. What does it mean? It seems to mean something like this: An argument (containing a group of statements) is a deduction if the statements in the argument guarantee a conclusion. A deduction guarantees that a conclusion is true whenever its premises are true. Conclusions also can validly follow from nonsense premises, just as one deduces that Euclid was human from these premises:

> Euclid was a mathematician in ancient Alexandria.
> All mathematicians in ancient Alexandria were human.

But one might just as well conclude that Euclid was a monkey from this:

Euclid was a mathematician in ancient Alexandria.
All mathematicians in ancient Alexandria were monkeys.

So what does it mean to say that something is proven? It must mean more than simply having a valid deduction; we can see from the last two arguments that valid deductions can be constructed from false and even nonsense premises. It means that we must show that the premises themselves must be true to make sense of the argument. For a conclusion to be accepted, it must come from a valid deduction with accepted premises, but the truth of those premises depends on the semantics of the words in the argument.

Take the following syllogism:

No kitten that loves fish is unteachable.
No kitten without a tail will play with a gorilla.
Kittens with whiskers always love fish.
No teachable kitten has green eyes.
Kittens that have no whiskers have no tails.

It came from a conversation one summer morning in 1875 between Charles Dodgson (Lewis Carroll) and his friend Lord Tennyson as they sat in an Oxford teahouse eating peach jam cakes and making logical nonsense riddles. Several such syllogisms appeared in the symbolic logic book that Lewis Carroll would eventually write.[1]

Sherlock Holmes himself would say that none of these statements is true; yet, if they were, he would agree that "no green-eyed kitten will play with a gorilla." The truth or falsity of a compound statement should be distinguished from validity. Logic does not take positions on facts or philosophical principles. It is concerned with whether a valid conclusion can be drawn from a valid argument. Apply logic to an argument, and all it does is check for conflicts between the premises and the conclusion.

In the fall of some year of my distant past, I taught a course called Plausible and Deductive Reasoning. The course explored how observations in mathematics build intuition, and how that intuition, in turn, influences ideas for conjectures that chance to be deductively proven. My joy of teaching is greatest at those moments when I feel bona fide sparks of communication between my students and me. Nichole was in the class along with five others, Susan, Scarlet, Tristan, and two other students who took the class in unplanned directions with surprising results. I began with a riddle posed by the earlier kitten syllogism.

To my astonishment, it took Nichole just a few minutes to conclude that no green-eyed kitten will play with a gorilla. Nichole seemed to have an "inner-marking" scheme for organizing syllogisms. I thought that perhaps she had seen this particular well-known syllogism before and had just memorized the answer. It was hard to believe that she could untangle such a mixed-up syllogism without a map. So I created another nonsense syllogism to present at the next meeting:

Atoms that are not radioactive are always unexcitable.
Heavy atoms have strong bonds.
Uranium is tasteless.
No radioactive atom is easy to swallow.
No atom that is not strong is tasteless.
All atoms are excitable, except uranium.

Again, it took her less than two minutes to conclude, "Heavy things are not easy to swallow." What extraordinary talent!

In class, we explored compound statements and rephrased them so they became conditional. For example, "No kitten that loves fish is unteachable" was first rephrased as "All kittens that love fish are teachable," which, in turn, became "If a kitten loves fish, then it is teachable." Each component of this last statement had a possibility of being true or false. "Loving fish" was being compared to "being teachable." It meant that truth values could be assigned to the components "kittens loving fish" and "teachability." A kitten could love fish or not love fish, just as it could be teachable or not.

The statements "No teachable kitten has green eyes" and "No

kitten that loves fish is unteachable" were rewritten as follows:

All kittens that love fish are teachable.
All teachable kittens have green eyes.

From these, we deduced the following:

All kittens that love fish have green eyes.

This last statement is deductively inescapable from the previous two; its truth is absolutely guaranteed if each of the two previous statements is true. It is simply part of the "algebra" of human reasoning. It says that "all x are z" because "all x are y" and "all y are z."

One day I gave Nichole a number theory problem that I myself did not know how to begin. With extraordinary risk and unprofessional advice, I hinted that the problem was easy. A few hours later, Nichole came to me with an exceptionally elegant solution. I had to find out how she was able to untangle syllogisms.

"How did you get the conclusion so fast?" I asked.

"I'm not sure," she answered.

"You're not sure? Something must have been going on in your mind when you were untangling the syllogism. Can you recall how you were thinking?"

"No. The answer seemed to have just come..." she began to say. Then, after a brief pause, she said, "Wait a minute. I think that I thought of the whole syllogism as a bunch of fractions and cancelled terms."

"Of course!" I said to myself, and immediately realized how she did it.

"How would you do it?" she asked me.

"Well, you know," I said. "I was never instructed in the rules of logic or the schemes by which one can untangle syllogisms. Rarely do mathematicians have formal schooling in logic, yet they seem to do just fine in the maze of complex inference."

The "kitten and unteachable fish syllogism" is far too complex for the normal mind to quickly conclude what it tries to reveal. The mind needs strategies to help it do its job. Statements that involve words such as *all, some,* or *none* can be illustrated by intersecting circles. The four possibilities below are called *Euler diagrams.*

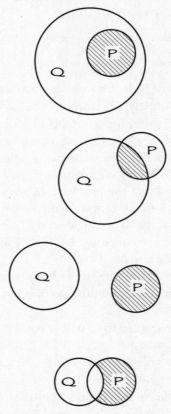

Henri Poincaré suggested that syllogisms can be arranged in such a way that Euler diagrams of nested statements can represent them. Perhaps logically valid arguments "feel" right because of this natural nesting.

A collection of statements connecting green eyes, kittens, tails, gorillas, whiskers, fish, and teachability could be too complex to untangle without a good method for organizing them.

Susan broke down the problem, statement by statement. Because every statement talked about kittens, she eliminated the word *kitten*

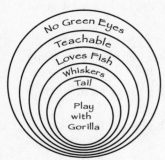

and abbreviated the attributes, and she thought of the attributes as simply names for the classes of kittens with such attributes. She put the name of the class in brackets ({ }), so {TEACHABLE} is the name of the class of *teachable kittens*; {LOVE FISH} is the name of the class of *kittens that love fish*, and so on.

With a little work, she could then construct the diagram of nested containment relations and untangle the syllogism.

Logic is used to prove theorems in mathematics. But mathematics is much more than simply logic. Perhaps we would all better understand mathematics if our arguments could be set in a clear nesting of syllogisms, such as the sort that's illustrated above. But that would surely reduce a highly intellectual branch of learning to a dull sport. We have seen that one can have a perfectly valid logical argument that produces nonsense or even untruthful facts. The principal question for us is, why do we understand anything? What gives us the very basic communication connections that guarantee understanding? How is it possible that a college freshman mathematics student who has never been taught the rules of logic or proof can unravel a syllogism and "feel" the truth of the conclusion? In other words, why is it that we know when we know?

Spoken and written languages are informal, and their ambiguities can trigger new ideas in the course of conversation. A spoken sentence might have several interpretations; and there are many ways of saying the same thing. An extreme example of language misunderstanding is found in a typically Carrollian dialog between Alice and the White Knight in *Through the Looking Glass*. The White Knight is about to sing a song to cheer up Alice, who seemed sad.

> "You are sad," the Knight said in an anxious tone: "let me sing you a song to comfort you."

"Is it very long?" Alice asked, for she had heard a good deal of poetry that day.

"It's long," said the Knight, "but it's very, *very* beautiful. Everybody that hears me sing it—either it brings the *tears* into their eyes, or else—"

"Or else what?" said Alice, for the Knight had made a sudden pause.

"Or else it doesn't, you know. The name of the song is called 'Haddocks' Eyes.'"

"Oh, that's the name of the song, is it?" Alice said, trying to feel interested.

"No, you don't understand," the Knight said, looking a little vexed. "That's what the name is *called*. The name really *is* 'The Aged Aged Man.'"

"Then I ought to have said 'That's what the *song* is called'?" Alice corrected herself.

"No you oughtn't: that's quite another thing! The *song* is called 'Ways and Means': but that's only what it's *called*, you know!"

"Well, what *is* the song, then?" said Alice, who was by this time completely bewildered.

"I was coming to that," the Knight said. "The song really *is* 'A-sitting On A Gate': and the tune's my own invention."

We are persuaded by reasoning, and reasoning involves the use of logic. But there is more to it—much more. First, it depends on the internal language we think in, which, in turn, involves words and sentences that transmit meaning from which we can gather either truth or falsehood. Take the example given by Wittgenstein[2] of different meanings of the word *is*. It could mean the copula, equality, or an expression of existence. One could say, "Green is green," ambiguously suggesting that we have either expressed a nonsense tautology, in which both the first and last words are simply names for the color green, or have meant the first word to be a proper name and the last to be the color green. Thus, Wittgenstein concluded, the only way to avoid confusion in relaying ideas is to form a logically precise language in which each word has a unique meaning. But we don't have a logically precise language, so our expressions fall prey to disagreements encouraged by environmental differences and cultural experiences. Wittgenstein observed that meaning is communicated by the sentence, not by the individual words.

You might be thinking that the language of mathematics is precise and, therefore, would qualify as a language in which each sign has a unique meaning. But mathematicians generally use an informal spoken language to communicate proof and ideas—their proofs are rarely put into a purely symbolic language form, so how is it that we assume their theorems to be universal and eternal?

Do we think in syllogisms? Many of my daily thoughts are "ordered" statements. I might say to myself, "I will wash the dishes after lunch," meaning "washing dishes follows having lunch." But thoughts can get much more complicated by webs of entangled ambiguities that cannot be systematized so easily. In 1854, George Boole wrote a book entitled *An Investigation of the Laws of Thought on Which Are Founded the Mathematical Theories of Logic and Probabilities*, in which he proposed to devise an algebraic system for modeling thought. It was, in fact, an ingenious system for mapping rules of deduction to rules of algebra. In a very naïve sense, we can metaphorically view much of common communication processing as a scheme for ordering thoughts by containment. So, the typical "if P, then Q" statement is viewed as a nest diagram with P contained in Q.[3] The project was up against strong opposition from the ambiguities of language (and, hence, of thought), but, as with many good ideas, it laid seeds for the growth of kindred ideas. Boole's project led to the development of symbolic logic, an algebra for sets, with rules and axioms that are similar to those of arithmetic. The old Aristotelian syllogistic logic was extended to a rich logic that could be manipulated by symbolic rules.

Boole stated five rules, one of which is that either a statement is true or its negation is true. His five basic rules of logic are instrumental in presenting convincing arguments and in governing clear thinking. But are they absolute? Our conventional logic dictates the use of *the law of the excluded middle*, which is the first rule on Boole's list. Could there be any logical sense to denying that something—say, "I went to the post office to buy stamps"—is either true or not true? If I did not go to the post office for no stamps, then didn't I go to buy at least one stamp? There is something so compelling about this rule of logic that it's hard to believe that there could be any other argument. But there is. In the early part of the twentieth century, several systems of logic, which included several possibilities for middle values between true and false, were developed. These are called non-Aristotelian logics.

Think of an argument you had with a highly intelligent, rational friend that ended in deadlock. You probably thought at the time that your viewpoint was totally logical, but it is likely that your friend thought the same about his. Could there have been a real conflict between the kinds of logic used? It's more likely that you had misunderstandings of the semantics. After all, you were conversing in a natural language with very flexible formal rules that might not have the strength and power to convey all feelings—one person's conception of pain, love, or sadness is quite different from another's. It is possible that you each came at the issue with a different debating strategy. Still, is it possible that you and your friend were using conflicting systems of logic? Is there more than one kind of logic? You might be surprised to hear that there is.

Are mathematical objects already there, independent of our thoughts, waiting to be discovered, or are they brought into existence by our own inventions? There is a school for each answer. Platonists believe that we discover mathematical objects and that all meaning comes from the relationship the objects have with one another. If nobody has seen the proof of a theorem in the mathematical forest, it can still be true. Such a school is forced to accept that proof of a theorem reduces to discovering its truth or falsity. On the other hand, the constructivist school sees a different kind of logic. Take the standard indirect proof that $\sqrt{2}$ is not rational. That proof assumes that $\sqrt{2}$ is rational, and that assumption leads to a contradiction.[4]

To prove that $\sqrt{2}$ is not rational, suppose that it is—that is, suppose that $\sqrt{2} = p/q$, a rational number with no common factors.

(If both p and q had common factors, they could be canceled.) Then square both sides to find that $2 = p^2/q^2$. So, $2q^2 = p^2$. This means that p^2 is an even number and, hence, that p is an even number. So, let $p = 2s$, for some s. Then $p^2 = 4s^2$ and, hence, $2q^2 = 4s^2$. Cancel the factors of 2 from both sides of the last equation to find that $q^2 = 2s^2$. This means that q^2 is an even number and, hence, that q is also an even number. This tells us that there are common factors of p and q. We therefore have a contradiction to what we supposed to be true, that $\sqrt{2}$ is a rational number.

In formal logic, we would then say that it is false to say that $\sqrt{2}$ is rational and, therefore, conclude that $\sqrt{2}$ is not rational. Constructivist logic will not accept such a method of proof. According to the constructivist school, all we can say is that the statement "$\sqrt{2}$ is rational" is not true. We cannot conclude from this that $\sqrt{2}$ is not rational.

Formal logic admits the law of the excluded middle, which, as we have seen, says that if statement A is false, then its negation must be true; likewise, if A is true, its negation is false. Formal logic has only two truth values, true and false. If the assumption of a truth value leads to a contradiction, the original truth value is switched from true to false or false to true. But what happens in a logical system with more than two truth values? Constructivists do not accept the law of the excluded middle and, therefore, do not accept the method of indirect proof, which permits a swapping of truth values whenever a contradiction is encountered.

It all comes down to the question of how your school views mathematical objects. Are they there waiting to be discovered, or are they brought into existence by invention? If they are waiting to be discovered, they must be true, independent of whether they are discovered; therefore, their negations must be false. However, if they are brought into existence by invention, the establishment of their truth depends on either an established collection of self-evident truths or not. In the latter case, truth or falsity makes no sense.

To complicate matters, Lewis Carroll playfully calls our attention to the problem of infinite regression of the hypothetical in *What the Tortoise Said to Achilles*.[5]

After Achilles and the tortoise finish a famous race—which we'll hear more about in Part 2—they rest while Achilles gives reasons for why his winning was possible. He explains that he was able to overtake his opponent because the distances in the infinite sequence of distances between them were diminishing. The tortoise thinks about this and, as a counterargument, describes a race course "that most people fancy they can get to the end of in two or three steps, though it *really* consists of an infinite number of distances, each one longer than the previous one." He takes an argument from the first proposition of Euclid while Achilles takes out an enormous notebook and pencil. "Proceed!" he says. "And speak slowly, please! *Shorthand* isn't invented yet!"

"Well, now," says the Tortoise, "let's take a little bit of the argument in that First Proposition—just two steps, and the conclusion drawn from them. Kindly enter them in your notebook. And in order to refer to them conveniently, let's call them *A*, *B*, and *Z*:

(A) Things that are equal to the same are equal to each other.
(B) The two sides of this Triangle are equal to the same.
(Z) The two sides of this Triangle are equal to each other."

While Achilles is busy writing in his notebook, the Tortoise explains that some readers, including himself, might accept *A* and *B* as true and still not accept the hypothetical proposition "*A* and *B* implies *Z*." In other words, to accept *Z* as true, one must also accept "If *A* and *B* are true, *Z* must be true."

The Tortoise asks, "What else have you got in [your notebook]?"

"Only a few memoranda," says Achilles, nervously fluttering the leaves, "a few memoranda of—of the battles in which I have distinguished myself!"

"Plenty of blank leaves, I see!" the Tortoise cheerily remarks. "We shall need them all!" (Achilles shudders.) "Now write as I dictate:

(A) Things that are equal to the same are equal to each other.
(B) The two sides of this Triangle are equal to the same.
(C) If *A* and *B* are true, then *Z* must be true.
(Z) The two sides of this Triangle are equal to each other."

Before long, Achilles is forced to enter this in his notebook:

(D) If *A* and *B* and *C* are true, then *Z* must be true.

Achilles protests that if one accepts *A* and *B* and *C* and *D*, then one must accept *Z* as true.

"Then Logic would take you by the throat, and force you to do it!' Achilles triumphantly replies. "Logic would tell you 'You can't help yourself. Now that you've accepted *A* and *B* and *C* and *D*, you must accept *Z*! So you've no choice, you see.'"

But the Tortoise does not buy it and insists that *D* would have to be granted as hypothetical, and so Achilles soon finds the leaves of his enormous notebook filling as he enters this:

(D) If *A* and *B* and *C* are true, then *Z* must be true.
(E) If *A* and *B* and *C* and *D* are true, then *Z* must be true.

Here the narrator, having pressing business at the bank, was obliged to leave the happy pair and did not again pass the spot until some months afterward. When he did so, Achilles was still seated on the bank of the much-enduring Tortoise and was writing in his notebook, which appeared to be nearly full. The Tortoise was saying, "Have you got that last step written down? Unless I've lost count, that makes a thousand and one. There are several millions more to come."

It's hard to argue that humans are programmed to be logical, when freshmen do so poorly on simple questions of logic. But the mechanics of thinking are far from the simple reasoning that declares *Z* to be true when we accept *A* and *B*, and "*A* and *B* implies *Z*." Linguists tell us that all languages have terms such as *not*, *and*, *or*, and *same*, and that in many parts of the world, children use *not*, *and*, and *or* before the average age of three. So there might be something to the idea that humans are quick to accept the basic building blocks of logic. In fact, it's not a far stretch to infer that humans might be programmed—or, at least, forced by their experience in communicating—to accept the modus ponens, the proposition of logic that tells us that *Z* is true whenever we know that *A* is true and that "*A* implies *Z*" is true. But the rules of thought extend much further than this simple force of logic.[6]

[handwritten margin note: refining the antecedent / law of detachment]

[handwritten note: notation: $P \to Q, P \vdash Q$]

[handwritten note: or $\dfrac{P \to Q, P}{Q}$]

[handwritten note: If x is true then y is true / x is true / Therefore y is true]

Twenty-three hundred years ago, mathematicians and philosophers struggled with the idea of defining terms such as *point*, *line*, and *circle*, mathematical objects that come from experiencing our spatial world. Euclid decided to define a point as "that which has no part." That's not a very satisfactory definition from a modern point of view, but surely one that is better than Plato's: "A point is that of which the middle covers the ends," whatever that means. (His definitions for a straight line and a circle are not much better.) Definitions of point and line are not necessary. If we postulate, as Aristotle did, that all men are mortal and that Socrates was a man, we are forced to conclude that Socrates was mortal, no matter what it means to be a man or mortal. The truthfulness of the conclusion rests on the truthfulness of the assumptions. This is the idea behind Euclid's postulates. They are true by virtue of assumption—they don't need to be proven. We can then conclude new truths, which follow logically from the assumptions.

Euclid's first proposition—that an equilateral triangle can be constructed on a finite straight line—also has a flaw, albeit one that doesn't seem to matter much. Only two postulates are used in the proof. They appear as Postulates 1 and 3 in Euclid's *Elements*.

1. *To draw a straight line from any point to any point.* By this, Euclid means that between any two points there is a unique straight line.[7]

3. *To describe a circle with any center and distance.* Here, the word *distance* can be interpreted as *radius*. Euclid was thinking that the circle consists of points that are all of the same distance from a fixed point, the center of the circle.

These are two of the five assumptions out of which Euclid built his *Elements*. They seem so self-evident and so basic to our experience with points and lines that proof is not necessary. Indeed, without some self-evident assumptions, we cannot have any notion of proof. If we accept them unconditionally, then, like the syllogism that claims Socrates mortal, these two postulates claim the truth of Euclid's first proposition. An equilateral triangle can be constructed on a finite straight line *AB*.

To prove it, Euclid had to use his third postulate to construct two circles of radius *AB*, one centered at *A* and the other at *B*. He then used his first postulate to mark one of the points of intersection as *P* and to construct two lines, one from *A* to *P* and the other from *B* to *P*.

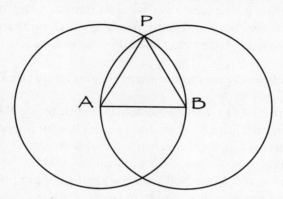

How do we know that the two circles actually intersect each other? If mathematics is axiomatic and a system of arguments that follow from each other, we should not accept Proposition 1 as valid, for there is no argument that tells us that the intersection of the two circles used in the proof exists. The problem is that we find it hard to believe that the intersection does not exist. For more than two thousand years, we felt confident that the proof of Proposition 1 was valid. It "feels" obvious that the two circles intersect. In this sense, we "feel" truth. We "know" that the proposition is true because we are so convinced, and the small detail of conceiving a possible way for the two circles to not intersect should not really get in the way of our believing the proposition.

Those circles are imagined as part of the real world of circles. However, in the nineteenth century, mathematicians realized that the real world does not matter. For example (and we'll hear more about this later), we now know that the fifth postulate of Euclid— which, in effect, says that through a given point not on a given line, there is one and only one parallel to the given line[8]—can be replaced by a postulate stating that "there are *no* straight lines parallel to a given straight line." This might sound false; but if you replace Euclid's fifth postulate with this new one, you have a new system that is logically consistent and one that gives a whole new geometry. The geometry is not that of the world we are used to, so our intuition plays a trick on us. We have a much harder time accepting statements in this new geometry and must resort to checking things not by "feelings" of truth, but by clear, logical inferences.

The symbols we use to communicate thoughts might call up the very different pictures.

"Can pictures be considered actual proofs, or are they merely used as symbolic icons?" I asked my math class for liberal arts students. "There is a well known theorem that says that the sum of the first n odd integers is equal to the square of n—that is, $1 + 3 + 5 + \ldots + (2n - 1) = n^2$," I said and then sketched the figure below. The picture is a perfect square built from right-angled collections of odd numbers of dots, with each angled collection fitting on the previous one.

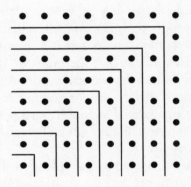

"Does the picture prove the theorem?" I asked.

"The picture is more convincing than the statement of the theorem," Thomas said as he moved his feet farther under his chair.

"Yes, but does it prove the theorem?" I asked again.

"I guess it doesn't really prove the theorem, but rather gives an impression that it is true because the picture is a square," Thomas said. "Maybe the picture helps us to begin to discover a real proof," he continued.

But other students disagreed. They saw the proof in the picture. So I presented another theorem. "If you add all the fractions $1/2^n$ together for every positive whole number n greater than 1, the result is just 1," I said while writing the equation

$$\frac{1}{2} + \frac{1}{4} + \frac{1}{8} + \cdots + \frac{1}{2^n} + \cdots = 1$$

I then sketched the picture (below) of a line of length 1 with divisions in increments of $1/2^n$ and asked whether it proved the theorem.

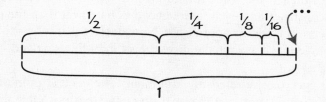

"The space left between the end of the line and the nth marker is getting smaller as n increases," one student said. "So, of course, there's nothing left by the time n gets to infinity." Everyone but Thomas agreed.

"No!" he said. "I don't believe it is possible to know that there will be nothing left in the space when n gets to infinity. I can't tell you why I don't believe it. I just don't."

"If anyone believes that the last picture proves that the sum of the fractions $1/2^n$ equals 1, then I have a picture that proves that 1 = 2," I said with a satirical smile and raised eyebrows. I drew a picture of an equilateral triangle with sides equal to one unit, then two equilateral triangles with sides half as long, and then four equilateral triangles with sides half as long again. The triangles in each picture were lined up, one on top of the other, so that the right side of the resulting figure was a straight line of length 1 and the left looked like the teeth of a saw.

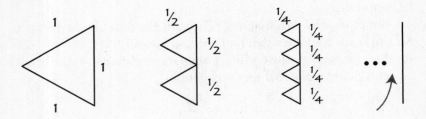

"Notice," I ventured, "that the length of the right side—the straight line side—of every one of these figures is 1, and that the length of the left side—the saw-tooth side—is always 2. As I continue my constructions, each figure becomes narrower than the previous one, so the widths approach 0. Now doesn't that tell us that the sawtooth side will eventually meet the straight line side and, therefore, that 1 = 2?"

But everyone rejected my argument. The general opinion was that one could not assume that the saw tooth would actually approach the straight line, even after an infinite number of iterations of the construction. The interesting thing here is that, although the majority of students were perfectly willing to accept the picture proof of the theorem that says that

$$\frac{1}{2} + \frac{1}{4} + \frac{1}{8} + \cdots + \frac{1}{2^n} + \cdots = 1$$

(reasoning that the remaining space shrinks to 0 as n grows large), all students refused to accept the picture proof that 2 = 1. Of course, one reason for the heightened suspicion is that we know that 2 is not equal to 1, and that knowledge psychologically conditions us to suspect that the picture is misleading. However, if the picture on page 67 implies the theorem that says that the sum of consecutive powers of 1/2 equals 1, my students should have accepted my picture proof that 2 = 1.

The second picture is persuasive but not convincing; we cannot tell whether the infinitely many small segments will fill the space or overshoot it, unless we examine the sum of fractions itself. Perhaps it was not fair to present pictures of infinite processes to my math for liberal arts class because one inevitably will fantasize, rather than see, what happens in the infinite tail of the picture.

This leads to the question of how to define *picture*. What do we mean by *picture*? What is the difference between the pictures on the left and the corresponding pictures on the right side of the illustration below? We have an intuitive understanding of what it means to add two numbers. Therefore, I think you will agree that the pictures on the left give the same mental impressions as those on the right. In fact, for small numbers, calculating on fingers is similar to mentally joining spots—the spots either represent fingers, or fingers represent spots. But the picture on the right seems to tell us something more than the symbols on the left: It gives us an impression that the sum of 1, 3, and 5 is a perfect square.

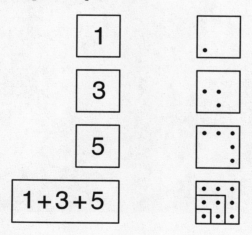

Interpretation of a painting banks on flexible meaning, interaction, history, and culture. What you see is what you interpret. Can you say what the Marrin Robinson painting represents? The answer depends on you, the viewer. Is it a woman at the beach, or the abstract female, or simply a study in forms and shades? The title might suggest what you see: It's called *Desert Prayer*. But if this picture were a mathematical object, there would be no flexibility in interpretation—what you see must lock into a fixed universal definition.

Desert Prayer, courtesy of Marrin Robinson, American University of Cairo.

CHAPTER 5

Legendre's Lament

Strange Worlds of Non-Euclidean Geometry

The wonderful thing about mathematics is that, in the end as well as in the beginning, it can depend upon no authority other than one's own (your own) mind; its verification comes from thinking alone, an activity open to anyone. If we have no theoretical equipment, we use the mathematical eyes and ears with which we were born and just experiment with our guesses to see whether we have faith in them.

—*Barry Mazur,* Imagining Numbers

On a winter afternoon in 1812, when Napoleon was on his way back from Russia by sleigh, having lost half a million unprepared Frenchmen to battles in the dark and frozen Russian winter, Joseph-Louis Lagrange went to the Académie des Sciences, where he was to give a lecture on Euclid's fifth postulate. This was a dark and gloomy time for France. It had come through twenty years of internal strife and foreign wars, when foreigners born in enemy countries were arrested and even beheaded. Lagrange himself was granted a special dispensation even though he was baptized Giuseppe Lodovico Lagrangia, a citizen of Turin, then the capital of the Duchy of Sardinia. Being Italian and living in France since his fifty-first birthday, through Robespierre's Reign of Terror, the revolutionary governments, and the Napoleonic wars, severely aggravated his nervousness, though just three years earlier Napoleon himself had named him to the Legion of Honor and Count of the

Empire. But Lagrange was confident about his mathematics and about the talk he was to give that day.

Just moments before his lecture, Lagrange was beginning to have doubts about his own proof. Something was wrong. He mumbled to himself in Italian as Adrien-Marie Legendre introduced him. Lagrange nervously stepped up to the podium. Though he was commonly recognized as the greatest living mathematician, he periodically suffered from bouts of profound melancholy, and something was clearly wrong at that moment. For an exceedingly long interval, he said nothing while fumbling with his notes. He looked up at his audience with his pale blue eyes. Then, in his usual feeble voice and Italian accent, he excused himself with a calm claim that he had found something wrong with his proof and walked off the stage, out of the hall.

The story of the fifth postulate reveals an important lesson for belief. The feeling of irrefutable truth is missing from it, and circular reasoning fooled everyone who tried to prove it. The turn of the eighteenth century, for example, also saw a glut of erroneous proofs of Euclid's fifth postulate. The real fifth postulate says this:

> If a straight line crossing two straight lines makes the interior angles on the same side less than two right angles, the two straight lines, if extended indefinitely, meet on that side on which the angles are less than two right angles.[1]

The picture below illustrates the postulate.

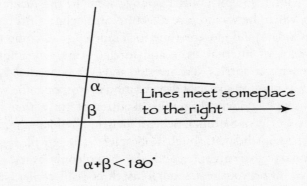

α

β

Lines meet someplace to the right ⟶

$\alpha + \beta < 180°$

The trouble with this axiom is that it is not as self-evident as any of the other four. It seems long and contrived, very unlike the simple fact that $1 + 1 = 2$, which everyone will agree to be self-evident. In other words, the feeling of irrefutable truth of the fifth postulate is missing, unlike the feeling of truth we have for the first postulate, which says that two points determine a unique line—namely, the one and only line that connects them. The first postulate is irrefutable, independent of experience or culture, as with the fact that $1 + 1 = 2$, but the fifth postulate has its own story.

Mathematical Compilation was the original title of a brilliant book on astronomy written in the second century A.D. It hypothesized an Earth-centered system of the universe and established astronomy as a science; it was the first work that attempted to explain the motions of the sun, moon, and planets from observations interpreted by applying mathematics. After several translations, its title became *The Greatest Compilation*. The Arabs translated it as *al-majisti*, but in Latin it became the *Almagest*. Its author was Claudius Ptolemy, a Roman-named Egyptian-Greek born in the first century. He was bothered by the fifth postulate and offered a very simple proof that goes like this:

In the figure below, assume that two lines, *AB* and *CD*, are parallel, and a third, *EF*, cuts across *AB* and *CD*.

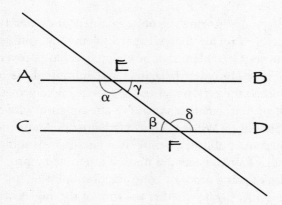

Then only one of three possibilities can occur:

1. $\alpha + \beta > \gamma + \delta$
2. $\alpha + \beta = \gamma + \delta$
3. $\alpha + \beta < \gamma + \delta$

Imagine the figure as if it is composed of two independent pairs of parallel lines (below). Then we can assume that whichever one of the previous possibilities occurs in one pair of parallel lines will occur in the other. So, $\alpha + \beta > 180°$ implies that $\gamma + \delta > 180°$. But this leads to the absurd conclusion that $\alpha + \beta + \gamma + \delta > 360°$. Clearly, the same absurd conclusion follows from the assumption that $\alpha + \beta < 180°$. So, the only possibility is that $\alpha + \beta = 180°$.

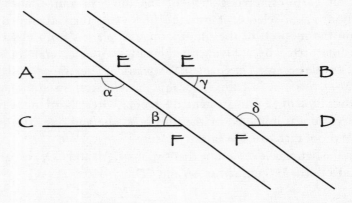

The fallacy in Ptolemy's proof remained undetected for three centuries before Proclus charged that Ptolemy had unintentionally assumed the very statement that he claimed to have proven. Proclus found Ptolemy's argument fallacious and submitted his own—but he, too, assumed the very statement he was trying to prove.

The list of mathematicians who've attempted to prove the fifth postulate reads like a Who's Who in mathematics. Many proofs were even accepted and published before their circular reasoning assumptions were discovered. Legendre himself attempted a proof with the brilliant idea of using proof by contradiction, splitting one hypothesis in three to investigate whether the sum of the angles of a triangle is equal to, greater than, or less than 180°. To contradict the second hypothesis, he had to assume that lines are infinite in length, something that is true for the only kind of geometry that was understood and accepted in the eighteenth century, but not for the new geometries that were discovered in the nineteenth century. Though he made many attempts to contradict the third hypothesis, he couldn't.

For thirty years, Legendre tried to prove Euclid's fifth postulate, publishing revision after revision of his popular version of Euclid's *Elements* with reorganized and simplified propositions. His *Elements* remained popular until the beginning of the twentieth century. He included an elegant proof of the following statement that was known a hundred years earlier: If there exists a single triangle having the sum of its angles equal to 180°, then the sum of the angles of *every* triangle is equal to 180°.

This proposition immediately suggested an alternative to the fifth postulate: There exists at least one triangle having the sum of its angles equal to 180°. In hindsight, we know that Legendre's proposition says nothing about what happens if there does not exist a single triangle having the sum of its angles equal to 180°. We know that in one non-Euclidean geometry, the sum of the angles of every triangle is greater than 180°, and in another, the sum of the angles of every triangle is less than 180°. The beauty of Legendre's replacement is this: To know which geometry you are in, all you have to do is pick any triangle and sum its angles.

Yet at each attempt to prove the fifth postulate, he found himself at every turn assuming statements that were equivalent to the fifth postulate. On and off for thirty years, he worked on proving the fifth postulate but failed. He simply would not (or could not) let go of a Euclidean point of view. He believed in it, and that belief blinded him to other possibilities. Even in his old age, when non-Euclidean geometries based on contradictory alternatives to the fifth postulate were emerging, he wrote this:

> It is nevertheless certain that the theorem on the sum of the three angles of the triangle should be considered one of those fundamental truths that are impossible to contest and that are an enduring example of mathematical certitude.[2]

A popular substitute for the fifth postulate was one that had been around since the fifth century and was revived by the Scotsman John Playfair. It says, "Through a given point not on a given line, one and only one line can be drawn parallel to the given line." This alternative has become known, fairly, as *Playfair's postulate*. Though there were many possible substitutes, Playfair's was the lovechild and the one that led Carl Freidrich Gauss, Johann Bolyai, and Nicholai Ivanovitch Lobachevsky to independently consider the possibility that through a given point not on a given line, there could be one line, no lines, or more than one line parallel to the given line. Remarkably, these alternatives don't lead to any contradictions—that is, the geometries that are described are entirely internally consistent. When mathematics is liberated from the sensations and perceptions of the observable world, it explodes into unlimited constructions and conceptions of the imagination. Upon discovering non-Euclidean geometries, Johann Bolyai wrote to his father, "I have discovered things so wonderful that I was astonished…out of nothing I have created a strange new world."

In hindsight, it is amazing to think that so many mathematicians, from Euclid to Gauss, were disturbed by the fifth postulate without doubting it. They were disturbed by unconscious inner feelings far stronger than their conscious arguments. They were disturbed because their inner feelings suggested that that fifth postulate was not self-evident. To let it go, however, would have contradicted everything their perceptions led them to *believe* about space.

The sudden freedom that comes from the insight that there might be more than one valid geometry not only alters the meaning of truth in mathematics, but also expands the world of mathematics itself. We learn that there are at least two (and possibly more) different geometries that are wholly consistent within themselves.

The fifth postulate has many equivalent alternates but Legendre's says that, "There exists at least one triangle having the sum of its angles equal to 180°." This gives a hint at why it is possible to have two contradictory geometries represent the real world. Has anyone ever seen a triangle that has the sum of its angles equal to 180°? The answer is: no! Outside of the mind, nobody has ever constructed the ideal triangle whose sum of angles equals exactly 180°. Even if we tried to construct one with absolute mathematical precision, we would have to ask, "How absolutely flat is the surface of our construction?"

The triangles we have experienced have contributed to our intuition of space; but we have not experienced all triangles. For example, we have not experienced enormously large triangles, just as we have not experienced the *chiliagon* (a polygon with one thousand sides). Could the sum of the angles of those enormously large triangles be less than 180°? Could it be that the sum of the angles of a triangle depends on the size of the triangle—the smaller the triangle is, the closer the sum is to 180°? That could mean that the triangles we humans normally see are so small that they are immeasurably close to 180°.

Bernhard Riemann was a student of Gauss, who considered the possibility that others of Euclid's postulates were interpreted too strictly. How should we interpret the postulate that says that a straight line is infinite? That's the one that says a finite straight line can be extended indefinitely in a straight line. Is that a self-evident truth? What could we mean by saying that something is infinite? Perhaps we mean that it never ends. Does *never-ending* have the same meaning as *infinite*? The Hatter in the Mad Tea Party would say, "Not the same thing a bit! Why you might just as well say that 'I see what I eat' is the same thing as 'I eat what I see!'" A circle could be viewed as a never-ending curve—one can move along it and never come to an end—yet it is not infinite in the usual sense of being boundless in space.

Okay, but one might say that a circle is not straight. What do we mean by *straight*? The usual definition of a *straight line* is the shortest path between two points. Riemann thought about this connection between straightness and distance and concluded that geometry itself is determined by the way we measure distance. As an example, consider constructing a triangle on Earth. We could draw the edges of the triangle as if they were passing right through Earth, or we could draw them directly on the surface of Earth.

In the illustration, the edges of the triangle are drawn as if the sphere is not really there and the corners of the triangle are just sitting on an imaginary sphere in the Euclidean space we normally experience. In the illustration on the right, each edge of the triangle is a path on the sphere that is the shortest distance between the two corners it connects. From where we stand in space, outside the sphere, those edges look curved, but as inhabitants who cannot see terribly far along a very large sphere, these lines would look straight and the world would seem Euclidean. A *great circle* on a sphere is defined to be the intersection of the sphere with a flat plane passing through the sphere's center. That intersection will be a circle in space that has the same radius as the sphere. Note that if lines on a sphere are interpreted as great circles, as they are in Riemann's geometry, the first postulate of Euclid's is no longer satisfied because any two points on opposite sides of the globe would have an infinite number of lines passing through them. However, the inhabitants of such a world would not see points on the opposite end of their globe.

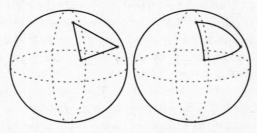

Euclid's propositions were about straight lines. In the *Elements*, a line is defined as something with "breadthless length." Such a definition does nothing, except to remind us, as Immanuel Kant said, that we already know what a line is. In Chapter 4 we saw that one could have legitimate logical arguments that yield nonsensical or even untruthful facts. The statement "Two startbingers determine a unique gobinger" seems to have no meaning; therefore, there would be no grounds to claim it to be true or false. However, if we replace the two meaningless words in the sentence with the words *points* and *lines*, respectively, we find that the statement suddenly makes perfect sense. Indeed, it is Euclid's first postulate.

Suppose one was to go through the thirteen books of Euclid replacing the words *point* and *line* wherever they are with the words *startbinger* and *gobinger*. Surely, Euclid's work would be just as true as it ever was. It makes one think that meaning must have crept in someplace. Then suppose that a person who had never studied geometry were to read and work though Euclid's *Elements*. Sooner or later (perhaps someplace around the tenth proposition), that person would begin to suspect that *startbinger* really means *point* and *gobinger* really means *line*. In other words, it is likely that real meaning would creep in by repeated experience of the words, and that the images of a startbinger and a gobinger in the reader's mind would be those of point and line in Euclidean space. However, Euclid does not use the fifth postulate before his twenty-ninth proposition, so an absolute meaning of a line is not rigidly set by logic until then. At Proposition 29, we have no choice but to think of a line as being the normal Euclidean straight line. If you never get past Proposition 28, your notion of line still has a chance at being something more general than simply the Euclidean notion of a straight line.

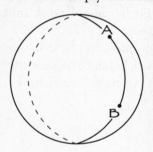

If you are on a curved surface such as Earth's, there are no straight lines in the Euclidean sense of "unbending" lines.[3] A straight line between two points on Earth must mean the shortest path between the points. Tokyo and San Francisco are both on the same 36° latitude, but the shortest distance between these cities is a path that passes close to the Aleutian Islands, nearly 50° latitude. Modern mathematics pays homage to the origins of geometry and Earth itself when it generalizes the notion of straight line to be defined as the path of shortest distance by calling the path a *geodesic*.[4] On a perfect sphere, geodesics are simply great circles.

Continue using the word *line*, but remember that what we mean is geodesic. So when we are talking about lines on the perfect sphere, we mean great circles. Notice that any two lines on a sphere meet at two points; parallel lines on the sphere do not exist. However, if you live on a large spherical planet and don't notice its curvature, you might think it is flat and might not notice that when you draw your lines, they bend ever so slightly toward each other.

A triangle in Riemann's geometry is three distinct points connected by three lines. With this more general concept of a triangle and its angles, we are ready to ask the question, what is the sum of the angles of a triangle? If the triangle is drawn on a flat plane, we know the answer: It's 180°. But if the triangle is drawn on a sphere, a quick sketch will show that the sum is greater than two right angles.[5] The figure on page 78 shows a Euclidean triangle and a Riemann triangle. Visual comparison shows that the sum of the angles of the Riemann triangle is greater than the sum of the angles of the Euclidean triangle. On a fixed sphere, the difference between the sums grows with the size of the triangle. This is just one of the strange (but logical) worlds opened up by changing Euclid's postulates.

Picture a disk.

First imagine that the disk is filled with points and that, for any two points A and B, the line passing through A and B is part of a circle that intersects the boundary of the disk at right angles. Except for this new definition of *line*, everything is Euclidean—the disk and all its points are sitting in a flat Euclidean space.

Next, imagine that every time you try to measure the distance between any two points A and B, your yardstick bends to become the unique circle passing through A and B that is perpendicular to the boundary of the disk.[6]

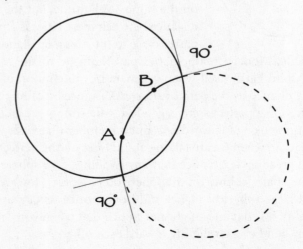

But that's not all: Your yardstick shrinks in such a way that, as you get to the edge of the disk, any point inside the disk is infinitely far from the boundary. So, you must imagine that the distance from *A* to *B* is the same as the distance between *B* and *C*, even though it doesn't appear to intuitively be.[7] Now this imagined space is certainly not Euclidean.

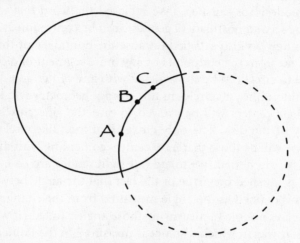

You have just imagined Poincaré's disk model for hyperbolic geometry, where the first four postulates of Euclid's geometry hold, but the fifth does not. Notice that for any geodesic *G* joining two points *A* and *B* and any point *C* not on that geodesic, there are infinitely many geodesics passing through *C* not meeting *G*.

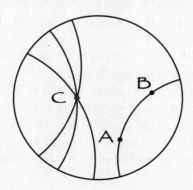

In Euclidean geometry, three points determine a unique circle. In the picture you were asked to imagine, there are only two points, A and B. But there is one added restriction: The circle passing through A and B must also meet the boundary of the Poincaré disk at an angle of 90°. That restriction determines the circle uniquely. So, the first postulate of Euclid's geometry (two points determine a unique geodesic) is satisfied. (We replaced the word *line* with *geodesic*.) The second postulate (a geodesic can be continuously extended indefinitely) is also satisfied because the boundary of the disk is infinitely far from the endpoints of any finite segment of a geodesic. The third (a circle can be constructed with any center and distance) is acceptable, although circles in this strange geometry will look like circles; their centers will be skewed toward the side closest to the boundary of the disk. The only circles that look like circles in the usual sense will be those that have centers coinciding with the center of the disk. The fourth postulate (all right angles are equal to one another) is satisfied by virtue of the fact that the angle between two geodesics is defined as the angle measured from their tangents, and distance plays no role in measuring those angles. In fact, if we ignore the peculiar way in which distance is measured in the Poincaré disk, every geodesic can be constructed from a line in the Euclidean plane by a projection through a sphere.

Suppose we were tiny specs living on a Poincaré disk of very large radius. We would not notice that our geodesics are curved, just as we don't notice that our Earth is curved when we look along its surface. But we would find geometry quite peculiar. For example, there would be no squares or rectangles because the sum of the angles of any quadrilateral would be less than 360°. We would also find that any two similar triangles (that is, triangles in which one is a magnification of the other) are congruent (that is, geometrically equivalent).

If the notion of straightness is connected to that of distance, then shouldn't we concern ourselves with what we mean by distance? Measuring distance is an old familiar idea; doesn't mathematics itself stem from the need to measure, to count, to survey land that we think we own? We think we know what it means to measure; surely all we do is pick a unit of measure, such as the inch, and mark off the number of inches from point *A* to point *B*. The ruler is the usual tool we use. But what if the unit of measure itself depended on where it is?

Four- and five-year-old children confuse distance with something connected to the time it takes to travel that distance. Ask a four-year-old "If you walk one mile from your house to school and return home after school, how far have you gone?" Chances are, the child will ask if the distance from school to home is also one mile. The child has to check that the distance going is the same as the distance back because time is confused with distance. A child might take more time coming home, kicking a stone or poking a stream with a stick. His or her impression of how long it takes to travel that distance gets confused with distance itself.

What a jolt! Kant said, in effect, that knowledge comes from the ability to perceive concepts from the sensations we receive, that space exists intuitively in the mind, that the postulates of geometry are imposed by the mind as *a priori* judgments, and that there could be no consistent reasoning about space without the postulates of Euclid.

Mathematics is not physics—it depends not on our sensations and perceptions of the observable world, as Kant had claimed, but rather on the conceptions of the imagination and the rigors of logic. For two millennia, we were concerned about the dependence of the fifth postulate, when all along we should have been celebrating the liberation of mathematics from the physical world.

Let's visit Playfair's version of the fifth postulate once again. "Through a given point not on a given line, one and only one line can be drawn parallel to the given line." Assuming it to not be true gives two possibilities:

1. No line can be drawn parallel to the given line.
2. More than one line can be drawn parallel to the given line.

(In the latter case, there would be infinitely many lines that can be drawn parallel to the given line.) The Riemann sphere (illustrated on page 79) is a model for the first possibility; the Poincaré disk is a model for the second.

Recall Legendre's version: "There exists at least one triangle having the sum of its angles equal to 180°." It, too, gives two possibilities:

1. There exists at least one triangle having the sum of its angles greater than 180°.
2. There exists at least one triangle having the sum of its angles less than 180°.

Again, the Riemann sphere is a model for the first possibility, and the Poincaré disk is a model for the second.

There are many alternatives to the fifth postulate. Here are just two more:

1. If there are three right angles in a quadrilateral (a four-sided polygon enclosing an area), the fourth is also a right angle.
2. There exist at least two non-congruent triangles that are scaled images of each other. (In the language of Euclidean geometry, this means that the two triangles are similar.)

To test which geometry you are in, just pick any quadrilateral with three right angles and check whether the fourth is also a right angle. Or, if you find a pair of triangles that are similar and not congruent, you are in Euclidean geometry.

It turns out that any one of the non-Euclidean geometries is just as valid as Euclidean geometry. (Details of why that is so are beyond the scope of this book.) It might seem that the non-Euclidean geometries are just abstractions that have nothing to do with the physical world we live in. Yet, Einstein's general theory of relativity generalizes Riemannian geometry to model space-time, suggesting that the universe is not Euclidean after all, but rather dimpled, with curvature smoothly changing from point to point, depending on mass size.

To determine which alternative to the fifth postulate best models the geometry of our world, we inevitably must test space by experimental observation. One such test is to measure the sum of the angles of a huge physical triangle in space. But we have a problem: To show that space is Euclidean, we would have to determine that the sum of the angles of our triangle precisely equals 180°. Because our instruments are never completely accurate, we would never know whether the sum of the angles of a physical triangle is 180° and, therefore, would never be able to physically demonstrate that our space is Euclidean. If our instruments were accurate to within, say, a millionth of a degree, all we could know is that the sum of the angles of our triangle is between 179.999999° and 180.000001°. We would need to know that it is precisely 180°. However, if the sum turned out to be 179°, we would undoubtedly know that we are in a hyperbolic space. So, to know that our universe is not Euclidean, our observed measurements must exceed the expected error of our instruments. To date, this has not happened.

This means that truth is no longer connected to what we see, hear, and experience. If I could invent an arbitrarily artificial collection of independent consistent postulates of geometry from the boundless imagination of my mind, I would have a geometry as true as Euclid's. Indeed, many exotic geometries, free of contradiction, have been invented, with counterintuitive theorems that say things such as "A line is perpendicular to itself." Strange as they are, the non-Euclidean geometries that were invented by Bolyai, Lobachevsky, Riemann, and others are no more or less true than Euclid's.

In the words of Henri Poincaré, "One geometry cannot be more true than another; it can only be more convenient."

Deductive logic might have originated in the West during the sixth century B.C., when Thales, Pythagoras, and their contemporaries started talking about large classes of objects without specifying particulars. Instead of just knowing that a triangle with sides 3 and 4 has hypotenuse 5, these early Greeks surmised that the length of the hypotenuse of *any* right triangle could be determined by a theorem that applies to all right triangles. In a relatively short period—less than three hundred years—a massive body of general statements about mathematical objects was organized in a gigantic web, each statement linked to others by invisible threads of logical consequence. The man best known for spinning this web is Euclid.[8] The logic used to prove such statements was not new, however. It must have been around since the beginning of language. In fact, it must have always been a part of language itself.

It is not surprising that Greek mathematicians encountered infinity and its surprises so soon after deductive logic was introduced as a method of arguing the truth of general statements. They found deductive arguments proving that there is no number that measures the magnitude of the diagonal of a square whose sides measure one unit. If the Pythagorean theorem is true, we accept that there are nonmeasurable things as simple as the diagonal of a square. It's hard to imagine physicists or engineers losing sleep over this. As Tobias Dantzig said in his marvelous book *Number*, "The fact that certain magnitudes, like $\sqrt{2}$, π, or e, are not expressible mathematically by means of rational numbers will not cause him [or her] to lose any sleep, as long as mathematics is furnishing him [or her] with rational approximations for such magnitudes to any accuracy he [or she] desires."[9] However, a different sort of logic was needed for mathematics to furnish rational approximations to any degree of accuracy, one that does not come with language. The logic of infinity is often counterintuitive and has many surprises, as we shall see.

PART II

Infinity

CHAPTER 6

Evan's Insight

Counting to Infinity

PROFESSOR: *You know how to count? How far can you count up to?*
PUPIL: *I can count to…to infinity.*
PROFESSOR: *That's not possible, miss.*
PUPIL: *Well then, let's say to sixteen.*
PROFESSOR: *That is enough. One must know one's limits.*

—Eugene Ionesco, The Lesson

In *Out of Africa*, Isak Dinesen talks about trying to learn numbers in Swahili from a Swedish dairyman speaking in Swedish.

As the Swahili word for nine, to Swedish ears, has a dubious ring, he did not like to tell it to me, and when he had counted: "seven, eight," he stopped, looked away, and said: "They have not got nine in Swahili."

"You mean," I said, "that they can only count as far as eight?"

"Oh, no," he said quickly. "They have got ten, eleven, twelve, and so on. But they have not got nine."

"Does that work?" I asked, wondering. "What do they do when they come to nineteen?"[1]

Does that work? There cannot be gaps, for otherwise there would be problems: What, other than *ten* could be the square of *three*? Clearly, the only possibility is to have *ten* replace *nine* and have it simply be a shift of words; *twenty* would replace *eighteen* (which would have otherwise been called *nineteen*); *thirty* would replace *twenty-seven* (which would have been *twenty-eight*, which is the real *twenty-nine*), and so on. "Dear me, it's so confusing!" Alice would say. Fortunately, the set of natural numbers is infinite, so each dubious-sounding number can be replaced by its successor.

Almost all of pure mathematics depends on the definition of a natural number. In turn, science and engineering rest on the correctness of pure mathematical models. How extraordinary it is, then, that we seem to understand numbers correctly from almost our first encounters with them. In the *Epinomis,* a little known and short dialogue of Plato, an Athenian argues:

> [A] creature's soul could surely never attain full virtue if the creature were without rational discourse, and that a creature that could not recognize two and three, odd and even, but was utterly unacquainted with number, could give no rational account of things whereof it had sensations and memories only, though there is nothing to keep it out of the rest of virtue, valor, and sobriety. But without true discourse a man will never become wise, and if he has not wisdom, the chiefest constituent of full virtue, he can never become perfectly good, and therefore not happy.[2]

Natural numbers had been used correctly for thousands of years before anyone thought of analyzing them. Only in the past century have we felt the need to find axioms for arithmetic. The superstructure of almost all of mathematics was not compromised by the lack of understanding of its core foundations. In fact, we feel comfortable using numbers before knowing what they really are.

Geometry has been examined and studied for more than two thousand years. But the axioms of arithmetic were not formulated before the beginning of the twentieth century. "Natural numbers" is the natural name for such numbers. We surely don't need axioms of arithmetic to know how to count. The *Epinomis* continues:

Well then, let us go on to face the real point we are to consider. How did we learn to count? How, I ask you, have we come to have the notions of one and two, the scheme of the universe endowing us with a native capacity for these notions? There are many other creatures whose native equipment does not so much as extend to the capacity to learn from our Father above how to count. But in our case, God, in the first place, constructed us with this faculty of understanding what is shown us, and then showed us the scene he still continues to show.[3]

To be sure, we can make abstract constructions of number systems and reduce properties of these number systems to a very small number of axioms. But the consequences that follow must lead to normal worlds that are consistent with the principles that we have established by design—namely, that, what Bertrand Russell once said should be true, "We want to have ten fingers and two eyes and one nose."

Why don't we run out of names and symbols for counting? It's one thing to have the name for a number and quite another to have a symbol for it. It all started somewhere in ancient Iraq, about five thousand years ago when the Sumerians invented a reasonable system of numbers and number symbols for trading goods. Fortunately, the Sumerians recorded their accounts on clay tablets, which survived long enough to be recorded by other civilizations. So we know something about the evolution of number symbols. It seems that finger counting is responsible for their design. If you went to market to buy a fish in Ur during Abraham's time, you would raise one finger as a signal that you wanted only one fish; you would raise two fingers if you wanted two fish. The orientation of your hand could be vertical or horizontal. Thus, the symbol for 2 could be designated by two vertical fingers or two horizontal fingers, which are rapidly sketched as two horizontal lines, which, in turn, are evolutionarily corrupted by faster and faster representations.

For most countries, the symbols for the first three numerals are either horizontal or vertical lines, most likely evolved from representations of fingers. When we reach the symbol for four, we generally do not see four vertical or four horizontal lines, but rather a configuration of four sticks. For some cultures, the transition from parallel line markings to other configurations does not occur before the number 6. The Chinese system (below) is one of the oldest, and one can see a logical finger- or stick-counting progression.[4]

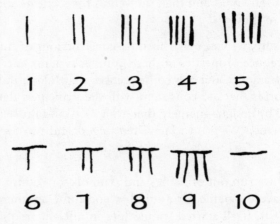

The symbol for 6 should not be six vertical sticks because it would be difficult to distinguish five vertical sticks from six without counting; the whole point of the symbol is not to have to count. This is very similar to contemporary number marking, in which five is marked with four vertical lines and a fifth horizontally crossing through.

It's almost as if the magic of symbol design and number naming is divinely inspired: We never run out of symbols or names for numbers. We need only 10 ten symbols—0, 1, 2, 3, 4, 5, 6, 7, 8, 9—to represent all numbers. For example, 427 really means[5] $4 \times 10^2 + 2 \times 10^1 + 7 \times 10^0$. But 427 is also equal to $1 \times 2^8 + 1 \times 2^7 + 0 \times 2^6 + 1 \times 2^5 + 0 \times 2^4 + 1 \times 2^3 + 0 \times 2^2 + 1 \times 2^1 + 1 \times 2^0$. If we understand that this number is constructed using powers of 2 instead of powers of 10, then all we need to know is the sequence of 0s and 1s to know the number. For example, the number we normally represent as 427 can also be represented as 110101011, as long as it is understood that each of the 0s and 1s must be multiplied by a power of 2.[6] In other words, 427 and 110101011 are just symbols that represent the same number, as long as we understand that

$$427 = 4 \times 10^2 + 2 \times 10^1 + 7 \times 10^0$$

and that

$$110101011 = 1 \times 2^8 + 1 \times 2^7 + 0 \times 2^6 + 1 \times 2^5 + 0 \times 2^4 + 1 \times 2^3 + 0 \times 2^2 + 1 \times 2^1 + 1 \times 2^0$$

The latter type of representation is called *binary*. If we see or think of the number 3 as the symbol 3, it registers a meaning connected with past events that trigger sensations associated with the number 3. Perhaps we make associations with the Three Musketeers or three turtledoves?[7] But 3 is just a symbol that might also represent 11—not the number eleven, but the binary code representation of the number 3. In the case of 3, we might mean it to be 3×10^0 in digital representation, or $1 \times 2^1 + 1 \times 2^0$ in binary representation, which would be represented as 11.

The ability to add and multiply must begin with some marking scheme, whether from counting fingers, stones, or something more imaginary. At first, counting was done concretely, by pointing to the objects one by one. Remnants of Aztec languages use numbers such as one stone, two stones, three stones, and so on. There are South Pacific languages that count one fruit, two fruit, and three fruit. In time, however, counting—in terms of such specific groups of objects as fingers, stones, fruit, and grains—developed to an abstract stage, in which the character of the objects being counted was no longer important. This was mathematics. The formation of the idea of number in the abstract sense developed as a result of repeated counting on fingers or by some other marking scheme.

"He still counts on his fingers," Timothy Bailey said in a concerned tone over the telephone. He was referring to Evan, a well-mannered eight-year-old boy with a vocabulary well beyond his age. Evan's fourth-grade teacher was calling me to ask if I knew any students who could help Evan learn his multiplication table.

"Why come to me? I teach college math and have very little experience teaching children fourth-grade math," I argued, thinking about my lack of training and expertise in such matters. Although I myself didn't feel particularly qualified to teach math to a young lad, I agreed to meet with him once a week to see what I could do.[8]

For two months, Evan came to my office after school once a week. At our first meeting, I gave him a sheet of twenty simple multiplication problems, using only one-digit numbers, such as 4×6 or 6×6, and asked him to complete the sheet in three minutes. It was an arbitrary amount of time—I had no idea how long it would or should take him. I pretended to be sifting through some papers, but all the while I watched to see if he was counting on his fingers. He was. By covertly using his fingers, he missed only three of the twenty problems.

We have strong evidence suggesting that all number systems evolved from counting fingers, toes, and other body parts. Children naturally use their fingers as the set into which they make a one-to-one correspondence with the names of numbers. Perhaps it is essential for arithmetic development. Yet some teachers still treat finger counting in the fourth grade as immature and shameful. Children still use their fingers, hiding them while they count, even though it is very likely that finger counting is a necessary stage for cognitive development of number concepts.

The Yupno, an Aboriginal tribe living in the remote highlands of New Guinea, count to thirty-three using an elaborate system that counts each finger in a given order and then counts body parts, alternating from one side to the other.[9] There is a definite advantage to the Yupno counting system. When American children count on their fingers, they start with a fist, raise each finger in succession, and stop at the final count. In the end, the number of fingers remaining in raised position is the answer. It presupposes no definite ordering: The child could start with any finger and raise any other finger that is not raised, though there are some cultural standards. Because the Yupno system requires counting in a definite order, it has an advantage: The answer is simply associated with the last body part in the count (as shown in illustration).

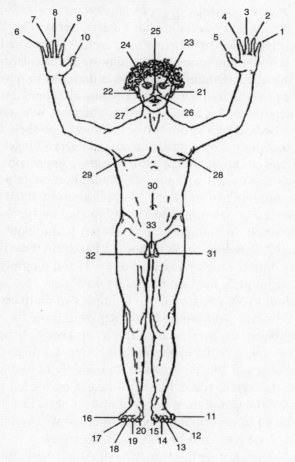

Yupno number system and counting. Wasserman and Dasen, *Journal of Cross-Cultural Psychology* (v. 25/1), 78-79, © 1994 Western Washington University. Reprinted by Permission of Corwin Press, Inc.

Back in the days when writing was—should I euphemistically say—inconvenient, finger reckoning was common. "The general purposes of digital notation were to aid in bargaining at the great international fairs with one whose language was not understood, to remember numbers in computing on the abacus, and to perform simple calculations."[10] The only complete record of ancient finger counting in existence is the codex *De computo vel loquela digitorum*, "On Calculating and Speaking with Fingers," written by Venerable Bede, an eighth-century Benedictine monk renowned among medieval scholars for, among other things, his calculation of the varying date of Easter Sunday, which was designed to never fall on the same day as the Jewish Passover. Because all other Church holidays are determined by Easter, Bede's calculations were considered significant. Bede illustrates how one can indicate numbers from 1 to 1 million by simply extending and bending fingers.[11] Finger notation leads to finger counting, which, in turn, leads to finger computation. Indeed, we don't need to know the multiplication table beyond 5 × 10 to multiply two numbers. Multiplication of small numbers can be reduced to counting fingers, multiplying by 10, and adding 100. For example, to multiply 6 by 8, subtract 5 from both numbers to get 1 and 3. Raise one finger on the left hand and three fingers on the right. Count the raised fingers (1 + 3 = 4) and multiply by 10 to get 40. Now multiply the bent fingers on each hand (4 × 2 = 8) and add the result to 40. You get 48.[12] To multiply two numbers between 11 and 15, subtract 10 from each, and represent those two numbers by raising fingers. Count the raised fingers and multiply by 10. Add the result to the product of the number of raised fingers on each hand, and then add 100. For example, to multiply 12 by 14, subtract 10 from each to get 2 and 4. Raise two fingers on the left hand and four on the right. Count the number of raised fingers (2 + 4 = 6) and multiply by 10 to get 60. Multiply the number of raised fingers on each hand (2 × 4 = 8). Add 100, 60, and 8 to get 168.[13]

Sixteenth-century texts show how this simple multiplication is carried out when writing is available.[14] They might even suggest the origin of the symbol for multiplication. If you want to multiply 6 and 8, form the *complements*, 4 and 2, by subtracting each number from 10. Now write the four numbers on the square grid, as pictured. To get the answer, 48, subtract 2 from 6 to get the 4, in the tens column. Then multiply the two numbers in the right column to get the 8.

In *What Counts: How Every Brain is Hardwired For Math*, Brian Butterworth asks the question, why is the left parietal lobe (the area of the brain where active movement of fingers is concentrated) also the area devoted to calculation?[15] Could it be that moving fingers is as necessary for counting as the eye is for seeing? If so, Butterworth's question—could it be that calculating ability comes from what we do with our fingers?—has an answer. His hypothesis is that it does. To come close to an answer, we need to put together several pieces of the finger puzzle. Wilder Penfield's famous mapping of the motor cortex (the part of the brain that controls motor functioning of the body) showed that the cells that control adjacent body parts are adjacent in the motor cortex.[16]

But there is more. Body parts that require more complex movement take up larger areas of the brain. Smaller body parts that require more complicated movements, such as the fingers, have a larger representation in the motor cortex than larger body parts that require less intricate movements, such as the arms. Another important consideration comes from extraordinarily surprising results of research with people who use Braille (and, hence, their fingers) to read. They have a larger motor cortex representation in the area serving the finger. Does the same phenomenon happen in the motor cortex of a person who plays the piano? A court stenographer?

Assuming that finger counting really does activate the motor cortex, which, in turn, increases the area of brain cells devoted to number thinking, is it possible that the reverse may also be true: if finger counting is repressed and no other "inner markings" come to the aid of thinking about calculations, that number thinking will not mature? In other words, is it possible that Evan's repressed "inner marking" schemes delayed his number thinking maturity?

Not knowing how to teach arithmetic to young children, I decided on a program to make Evan feel that his way of doing things might be different than what teachers expect and, at the same time, be no more or less right than any other way. At Evan's second visit, he brought several math homework problems to share. Some were very routine and ordinary, involving addition and subtraction in sequence. But one seemed out of place when we came across it. Here it is:

An archer has three shots at a target and hits it every time. Which of the following scores are possible? 7, 15, 16, 20, 24, 28, 29

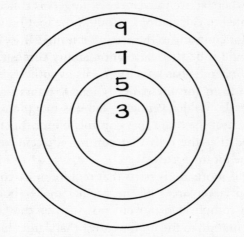

Evan rushed into the problem and started doing what I expect every boy in his class would do: add all possible groups of three numbers on the target. But one minute into his plan, without any word from me, he noticed that all the numbers on the target were odd and that there may be a clever way of getting the answer. For a moment, he didn't see how he could use the oddness of the target numbers, but he quickly noticed that some of the choices were also odd.

"Aha!" he said aloud. "Maybe this is a question about odd and even?" he asked.

My word of encouragement was simply a loud, "Hey!"

Evan was delighted with his thought. He asked if it was possible to get an even number by adding three odd numbers.

"If it isn't," he said joyfully, "then the answer can't be 16, 20, 24, or 28!"

I was impressed with his double-negative reasoning. I could have just told him that the sum of three odd numbers must be odd, but instead I told him to convince himself by adding several sets of three odd numbers. He did and said he was convinced.

"Well, you're easy to convince," I said, but added, "We'll investigate that later. Let's assume that when you add three odd numbers, you do get an odd number. What could you do with that information?"

Encouraged by his smart move, he immediately saw that he could eliminate 7 and 29 for different reasons. That left 15, which he saw was 5 + 5 + 5 and also 3 + 5 + 7.

My enthusiasm for his clever solution was so strong that I jumped on my chair. I lost my sense of decorum in the excitement of the moment.

"You are so clever!" I yelled. "Count on your fingers as much as you want!" I continued, jumping up and down.

How can I describe the smile on his face, the joy in his soul, the pleasure in his thoughts? Evan had found the keys to the problem. Certainly, for an eight-year-old boy, the concept of oddness and evenness is one of abstraction. He had made a connection between two sets of numbers, the target numbers and the answer choices. People often believe that mathematicians are good at calculations. In fact, often they are terrible calculators.

"Evan," I ventured to confess, "when I was your age, I also counted on my fingers. It's okay to count on your fingers. Some day you'll forget to count on your fingers and not notice that you stopped."

"Really?" he asked in amazement. "You used to count on your fingers?"

"Yes, I did," I answered. "Any pocket calculator can multiply," I continued, "and if that calculator had fingers, it would use them. The solution that you gave was elegant, not the kind that a calculator can give."

Evan felt triumphant. He sailed through the next exercise set of multiplication problems overtly using his fingers and getting a perfect score.

To be sure Evan understood and appreciated the beauty of his own idea, I asked him to write, in his own words, exactly what he did. Here is what he wrote.

Well we start off by knowing
that the numbers on target are
odd. If you add 3 odd numbers,
you get an odd number. But 20,
16, 24, and 28 are even numbers
so they can't be the score. This
leaves 15, 7 and 29. The archer hit
all three times eliminating 7
because it is too small. Twenty nine
is too big because its bigger than
the largest number added 3x. So
29 is too big, leaving 15 (which is possible
by having the archer hit the
5 3x). Fifteen is also possible by having
the archer hit 7, 3 and 5.

Handwritten text courtesy of Evan Johnson.

The art of solving problems involves making connections between what is known and what is not known. Connections might grow from other connections or show up in surprising places. The information we consciously collect from one experience subconsciously transforms into new information that can be used for new purposes. Knowing requires understanding, and understanding requires connections to other things that are known.

Evan knew about odd and even numbers from some past experience; perhaps he learned about them as trivial pieces of information that were never applied to anything. He knew that a whole number is either odd or even, but he didn't learn about evenness and oddness to solve the archery problem. All he knew was that a whole number is either odd or even, and that an even number is a whole number that is twice another whole number; this knowledge was stored, waiting to be used for no apparent purpose. Had Evan never been told about odd and even numbers, either he would have solved the problem by laboriously checking every possibility, or—this is a remote possibility that happens only for geniuses—he would have discovered the concept of even and odd numbers on his own.

The bags of information that we collect while bushwhacking though the mathematical forest are what we use to connect what we know with what we do not. Teaching is the art of leading one's pupils to discovery. That art includes relaying facts that might or might not immediately relate to anything important. But the hope is that some of those facts will become useful at some future time. Why do we need to know the multiplication table when pocket calculators can multiply for us? The simple answer is that we learn to recognize that some numbers are different from others by recognizing the numbers in the table. It could be very handy to know that some numbers are missing from the table, or that 64 is the square of 8 because it is one of the numbers on the diagonal. The more difficult answer is that we simply do not know how we use subconscious information that comes to us through conscious experience. Here's how Hume put it:

> Were ideas entirely loose and unconnected, chance alone would join them; and 'tis impossible the same simple ideas should fall regularly into complex ones (as they commonly do) without some bond of union among them, some associating quality, by which one idea naturally introduces another. [R]egard it as a gentle force, which commonly prevails, and is the cause why, among other things, languages so nearly correspond to each other; nature in a manner pointing out to every one those simple ideas, which are the most proper to be united into a complex one.[17]

And here is Francis Galton, the English scientist:

> When I am engaged in trying to think anything out, the process of doing so appears to me to be this: The ideas that lie at any moment within my full consciousness seem to attract of their own accord the most appropriate out of a number of other ideas that are lying close at hand, but imperfectly within the range of my consciousness. There seems to be a presence-chamber in my mind where full consciousness holds court, and where two or three ideas are at the same time in audience, and an ante-chamber full of

more or less allied ideas, which is situated just beyond the full ken of consciousness. Out of this ante-chamber the ideas most nearly allied to those in the presence-chamber appear to be summoned in a mechanically logical way, and to have their turn of audience.[18]

I cannot suggest that solving Evan's target problem by testing every possibility is not mathematics, but surely everyone will agree that Evan's solution is elegant. If you ask two different mathematicians what mathematics is, you will get two different answers, neither one of which will be satisfactory.

One might say that mathematics is the manifold forms of organizing unintelligible and inexpressible thought connections to understand and communicate them clearly. The "manifold forms" of mathematics admit the inelegant forms, as well as the elegant, when expressing a particular connection. If Evan had taken the route that tested every possible choice, it, too, would have been considered mathematics. But his way was far more elegant. Jacques Hadamard put it this way "[I]t is clear that no significant discovery or invention can take place without the will of finding. But with Poincaré, we see something else, the intervention of the sense of beauty playing its part as an indispensable means of finding."[19]

I posed a new question to Evan.

"What if the target keeps going with odd scores forever?" I asked. (I avoided the word *infinity*, feeling that it would confuse the question.) "Suppose the target went on forever. Could you ever get an even score with three hits?"

It took Evan a few minutes to think of an answer. Before answering, he asked an unexpected question.

"What do you mean by going on forever? It has to stop somewhere!"

"Why does it have to stop?" I asked.

"It has to stop somewhere," he repeated.

"It has to stop only if there is a largest odd number. Is there a largest odd number?" I asked.

"Well, no," he answered.

"How do you know?"

After a long pause, squinting his face, he said, "There will always be a next odd number because you can always add two to any odd number to get the next."

"Then why do you say that the target has to stop?" I asked.

"Well, where would you fit the numbers that go off forever?"

Evan managed to have a sense of infinity for numbers but not for space. This was a great surprise to me. I thought it would be the other way around for a fourth grader. Evan imagined his infinite line of numbers going off somewhere, perhaps crowding each other as they increased. His infinity was a bounded one.

"If the numbers increase forever why can't the target?" I asked once again. "Are you imagining an edge to space?"

"Is there an edge to space?"

"If I told you that there is no edge, could you then imagine the target going off forever?"

There was another long pause before he finally answered, "Makes no difference if it does or doesn't. He [the archer] is never going to hit the big numbers anyway. I mean, he's got to be a bad arrow shooter if he gets an arrow in the 59 circle."

What Evan did to solve the problem was to break the infinite collection of all integers into just two classes. His idea was a smart child's version of a mid-eighteenth-century creation inspired by the great Swiss mathematician Leonard Euler. Euler's radical idea was to collect numbers in sets arranged by their remainders after division by particular chosen numbers. For example, the set of all natural numbers can be thought of as being composed of two sets, the set of numbers that have no remainder after division by 2 and the set of numbers that have remainder of 1 after division by 2. On the surface, this doesn't seem to be any more useful an idea than simply saying that the set of natural numbers is the union of the set of even numbers and the set of odd numbers. But the ramifications are extraordinary: It reduces an infinite collection of things to a finite collection.

It was a rudimentary beginning of a concept that plays one of the greatest central roles in all of mathematics, *the theory of groups.*

One of the most basic objects in mathematics is the set of integers (all whole numbers, including 0, together with the negative whole numbers). In its most general form, it is simply a set of things with order, structure, and properties. Its structure is the operations of arithmetic that can be performed among its members. Addition is an operation that gives a unique integer for every pair of integers.

Let's go back to the idea of breaking the integers into two sets, the set of even integers and the set of odd integers. The set of integers is infinite. But if we are simply interested in a property of being even or odd, we could group integers according to their remainder after division by 2. Imagine that all even integers are placed into a bag labeled EVEN and all odd integers are placed into a bag labeled ODD. If we add any two integers from EVEN, we get an integer in the same bag. This is just another way of saying that the sum of two even integers is an even integer. If we add any two integers from the ODD bag, we get an integer in the EVEN bag. This is because the sum of two odd integers is an even integer. Finally, if we add an integer in the EVEN bag to an integer in the ODD bag, we get an integer in the ODD bag. Now we can do something that might seem very peculiar: We describe a way to "add bags."

$$\text{EVEN} + \text{EVEN} = \text{EVEN}$$
$$\text{ODD} + \text{ODD} = \text{EVEN}$$
$$\text{EVEN} + \text{ODD} = \text{ODD}$$

The addition table below illustrates the idea.

+	even	odd
even	even	odd
odd	odd	even

Now notice that we could have labeled the bags anything we wished, and the addition would have been the same. We could have called the bags Tweedledee and Tweedledum, or White and Red, or 0 and 1, and still the addition would have been the same.

+	white	red
white	white	red
red	red	white

+	O	1
O	O	1
1	1	O

This "bag arithmetic" inherits many of the structural properties of the integers. For example, $x + y = x + z$ implies $y = z$ when x, y, and z are integers (the cancellation law), but the same is true when x, y, and z are the labels of "bags." It turns out that if you have any other set with only two members that have these additive structural properties, it must have the same addition table. Mathematically, the two sets would be the same. The only difference is in the names. Mathematicians call this the *group* "Zee-mod-two," and denote it as $Z/2$. The letter Z denotes the integers (from the German word *zahl*, meaning "number"); the 2 denotes the modulus—all numbers divisible by 2 are equivalent, and all integers having remainder of 1 after division by 2 are equivalent. In other words, the integers are grouped into two classes: those equivalent to 0 and those equivalent to 1.

This kind of grouping is used in Euclidean geometry where congruent triangles are not distinguished from one another. In Euclidean geometry, any statement that is true for one triangle is also true for any other congruent triangle. To put it another way, suppose that you want to research traffic flow over a bridge. It's likely that you would want to ignore the makes and colors of all passing vehicles, but you might want to distinguish vehicles by weight. In other words, you would consider a red Ford Taurus and a silver Jaguar XK8 to be equivalent. Surely, the two cars are not equal—one is three times the price of the other—but they are equivalent for the purposes of your study. When Evan hit on the idea that he had to distinguish only between oddness and evenness, he realized that he should make no distinction among 16, 20, 24, or 28. That enabled him to reduce a large number of trials down to just two.

Several months after Evan's wonderful solution to the target problem, I found him doodling a drawing that looked like a Jordan curve. His picture was immensely complicated, so I asked if he could find a way to discover if any random point was inside or outside his curve. To show him what I meant, I randomly picked a point and asked for the best scheme for telling if a point is inside or outside the curve. It took Evan less than three minutes to have a clear and sure

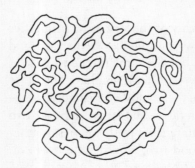

way of telling. "Draw a line from the point," he said, pointing to the point and drawing a line from it toward the southeast corner of the paper. "And count the number of times this line crosses the curve. If the number is odd, then the point is inside. If the number is even, the point is outside." He said this without hesitation. He didn't even have to check whether inside meant an odd number of crossings or an even number.[20] Once again, Evan subliminally connected his past experiences of oddness and evenness to a problem that would have otherwise been tedious to solve.

"Do you remember the problem with the archer shooting arrows at the target of odd numbers?" I asked.

"Um, yes."

"Do you remember the target that went on forever?"

After a long hesitation he answered, "Yes."

"Take the drawing you just did—the doodling you just did—and suppose it went off in all directions forever," I said, extending my hands far to my sides.

"I can't draw that!" he exclaimed.

"Only suppose. I know you can't draw it—nobody but God can—but suppose it was there when you came. There it was, a drawing from God."

"Okay," he said indulgingly.

"Okay, got it? Now, how can you tell if a point on the never-ending sheet of paper is inside the curve or outside?"

"That's silly!" he yelled. "There is no outside!" There was a long pause while he smiled. Then his smile broadened and his eyes widened while he raised his index finger to say, "The only thing on the outside is God."

CHAPTER 7

Encounters on the Aegean

Where the Finite Meets the Infinite

He thought he saw an angel
Dancing on a pin:
He looked again, and found it was
The shadow of a twin:
"If two fit on the head," he said,
"Why can't a third fit in?"

He thought he saw infinity
As something to amend:
He looked again and saw it needed
Plenty to append:
"And now I ask you this," he said,
"How could it ever end?"

—J.M.

P oseidon's fresh breath might have filled the sails of a thousand ships to enrich Aegean history, but color is what gives that ocean its matchless character, the blue of sapphire. Porpoises play in their own galaxy of white flecks on blue waters under skies swept clear by etesian winds flowing down from Russia. I was there when few Greek islands had airports, and ferries followed puzzling zigzag routes with wildly broken schedules. Ancient Greeks must have been about as puzzled when they first encountered infinity in the fifth century B.C. My first musings on the subject came while sailing in the vicinity of Pythagoras's birthplace. The year was 1963.

With no particular destination in mind, I boarded the first ferry leaving Piraeus. It made an odyssey northward against the *Meltimi*, stopping at Salonika, Alexandroupolis, and finally Lemnos, an island on the way to nowhere in the northern Aegean, keeping the magnificent Mount Athos within sight just until the Turkish Mount Ida could take over. Off the main shipping route with no airport, Lemnos had few visitors, except for the occasional cruise ship or rich American who came by yacht. This was the home of Hephaestus, god of the anvil, and the island where, according to myth, women killed their husbands. Jason and the Argonauts found it to be a paradise of solitary women.

From my hotel at the port of Castro, I noticed an attractive woman, not a local, walking in the marketplace from stall to stall buying fresh figs and olives, swinging her net bag and flirting with vendors to bargain. The next day, playing with my empty demitasse of Turkish coffee at a café in the harbor, the scent of lemon in the air, I watched the hem of her white cotton dress lash her tan knees left and right as she walked across the quay. She rested cross-legged at a capstan on the wharf.

"You didn't buy any fruit today?" I shouted in English. She turned with a smile. A little thrown, she cocked her head to one side and let out a delicate giggle. "The plums are not ripe yet," she replied. She continued her smile as she approached my table.

"Won't you join me in a morning coffee?" I asked with a hand inviting her to sit down.

"No, thank you. Don't drink coffee," she said, approaching, swinging her empty net bag. "Buy me a lemonade."

Fredericka was Swedish-Greek. Until then I thought all Swedes were blond. Fredericka had sun-bleached auburn hair that hung in rivulets of curls draped in front and back of the olive skin of her shoulders, and green eyes.

To my delight and astonishment, without much more conversation, she invited me to a yacht anchored in the harbor.

A wealthy Norwegian amateur archeologist by the name of Carl Hambro, who devoted much of his life to exploring the ruins of ancient civilizations, owned the yacht, *Hydra*, a black hulled, two-mast sailboat with white sails and a formidable bowsprit. She had an impressive gunroom and library, as well as a formal dining room.

Though he never completed his Ph.D. in the subject, Carl fancied himself a scholar of Schliemann's excavations of Troy. He had just completed a tour of ancient sites along the Turkish coast, visiting Melitos, Troy and a few important islands off the Ionian coast. Speaking an Oxford dialect of modern Greek—and, very likely a scholarly dialect of modern Turkish—he was interviewing locals who still claimed descent from Agamemnon for his thesis. Carl acquired his wealth from the simple invention of the tea bag. It was an innovation that sold well to American consumers but was frowned on and lampooned in Britain. His English grandfather patented the idea back in the nineteenth century and invested wisely in a small tea company by the name of Typhoo.

After Fredericka told Carl that I was studying mathematics in Paris, he brought me to his small library to find a book and offer me a job: tutoring math for free food and a luxurious tour of the Greek islands and Turkish coast. The book was an old classic called *Calculus Made Easy*, by Silvanus Thompson, possibly the slimmest calculus text ever printed. Penciled notes filled its margins. It seemed clear to me that Carl had been studying calculus on his own for many years, with increasing feelings of uncertainty over the subject. I had not taught anyone anything before and felt strangely uncomfortable with the idea, but, as I said before, I had no other destination in mind. I agreed to take the job.

His first question threw me. "Prove that there are infinitely many primes," he said looking a full ninety degrees away from me. His question's loose connection to calculus gave me the impression that he was testing my knowledge, but it just happened that I had recently seen a proof on the subject.

I said, "Suppose there are only a finite number of primes—say, k of them, with the largest one being P. So, we have a finite list of primes from 2 to P. Add 1 to the product of the k primes to get $(2 \cdot 3 \cdot 5 \cdot \ldots \cdot P) + 1$. Now I claim this new number—call it N—is either itself a prime or it is divisible by a prime greater than P."

"I'm listening," Carl replied.

"Suppose N is not a prime. Then $N = (2 \cdot 3 \cdot 5 \cdot \ldots \cdot P) + 1$ must be divisible by some prime number q."

"I don't see why."

His petulant tone dampened my ability to think logically.

"Because any number can be uniquely expressed as a product of primes. That's what the Fundamental Theorem of Arithmetic tells us."

"But you haven't proven that Fundamental Theorem yet."

"Sorry. How far back do you want me to go?" I asked.

"As far back as it takes for me to fully buy your proof."

"Okay, okay. Let's assume that the Fundamental Theorem is proven for the moment. I'll try to prove it later. But if we assume that the Fundamental Theorem is true, then you must see that N is divisible by some prime number q. Yes?"

"Yes."

I was beginning to sense ancient bearded heroes and philosophers listening at our side as the yacht plied forward through the white specs on blue water that came and quickly vanished. I found myself thinking ahead, not only of the proof, but also of my choice of words. Euclid proved this theorem more than two millennia ago and seemed to be listening to make sure I was getting it right.

"I claim that q is not any of the primes on our list. Why? Because if q were one of those primes, it would divide both N and $2 \cdot 3 \cdot 5 \cdot \ldots \cdot P$, and that would mean it would divide the difference $N - (2 \cdot 3 \cdot 5 \cdot \ldots \cdot P)$. But that difference equals 1 and...."

"Hold on," Carl said, "Why must it divide the difference?"

"If a number q divides each of two numbers A and B, then it also divides $A - B$."

"Why?"

His interruptions were reasonable, but at the time they threw my concentration off track. Some things you just know are true and don't think much about why. It wasn't difficult to demonstrate that if a number divides two numbers, then it also divides their difference, so I answered his question.

"It divides A, so there is some whole number s such that $A = q \cdot s$. That's what it means for q to divide A. In other words, A is a multiple of q. Likewise, $B = q \cdot r$, for some whole number r. So, $A - B = q \cdot s - q \cdot r = q \cdot (s - r)$. This shows that $A - B$ is a multiple of q and, hence, that q divides $A - B$."

"I see. So, because q divides both N and $(2 \cdot 3 \cdot 5 \cdot \ldots \cdot P)$, we know that it divides the difference $N - (2 \cdot 3 \cdot 5 \cdot \ldots \cdot P)$, which happens to equal 1."

"Yes."

"But how can the prime number q divide 1 without being equal to 1?" Carl asked.

"Exactly," I said. "It can't divide 1 without being equal to 1."

"I see. Because q is a prime number, it must be larger than 1. But it can't be larger than 1 because it must also be equal to 1. We have a contradiction."

"That's right. The only assumption we made was that there were a finite number of primes. The contradiction implies that that assumption was wrong. Therefore, there are infinitely many primes."

"That leaves the Fundamental Theorem of Arithmetic. But save that for tomorrow."

Phew, I made it through the first lesson without disaster.

The next morning, the light of a buoyed red sun beaming through my cabin window awakened me. I came on deck to find Fredericka having breakfast with Carl, listening to a jukebox playing Manos Kadjidakis songs. Yes! A Chicago jukebox, complete with cursive lighting, right on deck. Fredericka later told me that Carl had bought the machine from a café in Solonika and paid top dollar for it so the owner could not refuse to sell it, and that it would probably be thrown overboard when a new toy struck his fancy. He did outrageous things like that—like the time he bought a small publishing company just to publish one of his forever-rejected books. He had no interest in the business, so it went bankrupt and his book was poorly reviewed. Years later, I came across his book at a sidewalk vendor's table in midtown Manhattan. I bought it for two dollars but couldn't understand the first page. I tried a few more pages before giving it back to the same street vendor. Carl had dark features and a constant smile sieved through a black moustache over his motionless mouth. Fredericka took almost no further notice of me.

"Good morning," Carl called out with a waving gesture to sit down. "We are headed to Mytilene, my favorite Aegean island," he said. He was referring to Lesbos. "Wait till you taste the olives of

Mytilene," he said while picking up an olive from a jar in front of him. "They are the best. 'The whole Mediterranean,'" he said quoting from a Lawrence Durrell novel while spitting a pit into the palm of his hand, "'the sculpture, the palms, the gold beads, the bearded heroes, the wine, the ideas, the ship, the moonlight, the winged gorgons, the bronze men, the philosophers—all of it seems to rise in the sour, pungent taste of these black olives between the teeth.' And the sweet chestnuts, too, are out of this world!" He was a literate man who saw archeology as only one of his fields, but he never completed a degree in any one of them.

I sat down to a typical Greek breakfast of butter and honey on a slab of bread. The ship's crew inconspicuously managed everything from navigation to meals. It seemed that Carl had no interest in the mechanics of sailing. He simply informed his crew of his destination and, perhaps, a weekly menu.

"What about that Fundamental Theorem you promised me?" he asked spitting out the pit of another olive.

"What? Now?" I asked.

"Sure, you know I won't believe there's an infinitude of primes until you fulfill your promise of showing me how you prove your Fundamental Theorem," he said with a wry smile.

"Okay!" I nervously agreed, feeling him mock me. But, little did he know, I was up half the night with thoughts alternating between a proof of such a theorem and fantasies of romantic adventures with Fredericka.

"So I am to prove to you that every positive whole number greater than 1 either is a prime or can be uniquely expressed as a product of primes?" I asked.

"I think that is what we agreed the theorem to be."

"And when I say unique, I mean that the primes themselves are unique and not the order in which they appear in the product."

"That's what I understand," he replied getting into the rhythm of a Socratic dialogue.

"Okay, suppose that n is a positive whole number. Then it is either prime or composite. If it is prime, then there is nothing to prove. So suppose n is composite."

"By composite, I suppose you mean not prime?" Carl asked.

"Yes. Now because n is composite, it factors into two smaller numbers, one of which is prime. In other words, $n = p_1 \cdot n_1$, where p_1 is prime."

"Why is that?" he wanted to know.

Fredericka moved to a nearby chaise to read a glossy Greek fashion magazine. Maybe the sound of lofty math-talk would impress her. Probably not. Her bikini revealed a tan torso and belly tightened by regular exercise. Carl and I continued.

"Pick, from the collection of all the numbers that divide n, the smallest one and call it p_1. Then p_1 must be prime, for otherwise it, in turn, would have a divisor that would also divide n and, therefore, be a smaller divisor of n. But we assumed that p_1 was the smallest divisor of n."

"Okay."

"So we have $n = p_1 \cdot n_1$, with p_1 being prime. Now use the same argument on n_1 to get $n = p_1 \cdot p_2 \cdot n_2$. Repeat the argument enough times to exhaust any remaining composite factors to get $n = p_1 \cdot p_2 \cdot p_3 \cdot \ldots \cdot p_k$."

"Good! But how do we know that these prime factors are unique?"

"Ah! Suppose there were two sets of factors—say, $n = p_1 \cdot p_2 \cdot p_3 \cdot \ldots \cdot p_s$ and $n = q_1 \cdot q_2 \cdot q_3 \cdot \ldots \cdot q_t$, with $s < t$."

"And you are assuming that all the ps and qs are prime?"

"Yes. And further, we can suppose that we have arranged the primes in increasing order of size."

"Why can you suppose that?"

I was beginning to lose track of what I wanted to say. Fredericka chose another song from the jukebox—Louis Armstrong singing "Blueberry Hill." Carl repeated the question and, turning to Fredericka, asked her to turn down the volume.

"Because the theorem claims that only the primes are unique, not the order in which those primes appear in the product. Now p_1 divides $q_1 \cdot q_2 \cdot q_3 \cdot \ldots \cdot q_t$ and, hence, it must divide q_k for some k. But q_k is prime, so its only divisors are itself and 1. Because p_1 is not 1, it must be q_k. Hence, $p_1 = q_k$."

"Aha! I get it. That means $p_1 \geq q_1$ because the primes are ordered by size. And a similar argument, starting with the qs instead of the ps, shows that $q_1 \geq p_1$ and, therefore, that $p_1 = q_1$. So we can reduce

$$p_1 \cdot p_2 \cdot p_3 \cdot \cdots \cdot p_s \cdot = q_1 \cdot q_2 \cdot q_3 \cdot \cdots \cdot q_t$$

to

$$p_2 \cdot p_3 \cdot \cdots \cdot p_s = q_2 \cdot q_3 \cdot \cdots \cdot q_t$$

Then we repeat the argument to show that all the ps are really the qs."

Carl seemed to notice he didn't have my full attention. "But what if you end up with more qs than ps?"

"That's impossible," I said.

"Why?"

"Well…. Well, then you would have 1 on the left side of your equation and a product of primes on the other. That's not possible!"

"Aha! Congratulations. I'm a hard man to convince, but you have earned your fare."

Louis Armstrong started singing "I Can't Give You Anything but Love."

Over the next few days, Carl brought me several papers that he had written for publication in a physics journal. They were rejected. He tried to explain them to me, but I was thoroughly puzzled by their rather spiritual nature. I was not the right person to judge, but I thought at the time that his papers were about sheer nonsense.

We called on a few tiny islands, some no larger than a few kilometers from end to end. When we came to the tiny island Agios Eustratius, Carl anchored the *Hydra* so we could take a swim at the edge of golden cliffs that separated the sapphire blue ocean from the turquoise blue sky above. Fredericka swam nude. She handed me a snorkel, a mask, fins, and a spear gun. She was ready to hunt for octopus. She and I swam together to a small cove around the bend from the ship's anchor. With my spear gun cocked, I followed close to her side, keeping my head down in the frighteningly silent water.

Two large fish darted between groups of barnacled rocks and waving seaweed, chasing a school of mackerel, drawing my attention to a small cave just below the water's surface. Lost in a feeling of both fear and excitement, I encountered my first experience of absolute silence and intrusion into an ichthyic underwater world. It was eerie; here, even the tiny mackerel seemed fearsome. When I saw a small cuttlefish defensively wiggling its ten tentacles in my direction, I froze until it concluded that I was inconsequential and propelled itself out of sight. Numbed by the encounter, and somewhat comforted by my powerful spear gun's offer of implied protection, I was compelled to explore further into the mysterious caves with my forefinger nervously curled on the trigger. Fantasies ran wild. I didn't dare to turn around, fearing a giant squid stealthily following close behind. Thoroughly absorbed by the experience, I wandered too far from Fredericka.

I lost her; even when I lifted my head above the surface, she was nowhere to be found. Panicking, I called her name and swam in circles, diving deep beneath the surface until I ran out of breath. I decided it would be better to quietly use my snorkel to search for her. Just then I felt the tentacle of an octopus surround my throat. I dropped the gun, grabbed the limb with both hands, and turned around. It was Fredericka, who quietly came from behind to whip the dead octopus around me. I panicked but quickly realized that she was pulling me closer for a kiss on the lips. Her naked breasts pressed against my chest. She must have felt the beat of my racing heart. I had never seen anyone more beautiful. I no longer felt the octopus limb around my neck, though it was still there. Until that moment, I had assumed that she was Carl's girl and thought I had no chance.

Staying aboard the *Hydra* would have led to trouble; I should have been forewarned by the character that Jorge Luis Borges calls "the marshy monster that becomes a prefiguration or symbol of geometric progression." Carl would not have taken kindly to Fredericka's flirtations. He had several impressive guns aboard. Imagined as bow chasers, he would shoot them at nothing, for no apparent reason other than the joy of earsplitting noise. Twice I saw him, when seas were rough, strip off his clothes astride the bowsprit, like Odysseus approaching Troy, brandish his Browning automatic as if riding a stallion bareback in hunt of mythical beasts. That second bowsprit mount determined the end of my short tutoring job. I decided to get off the yacht as soon as it reached the Bay of Yera at the southeastern corner of Lesbos. The *Hydra* managed to squeeze by the narrow opening of the bay to anchor near the town of Perama, a tiny dreamlike village with a peaceful view of the skyline of Mount Olympus across the Aegean. Olive trees dominated the landscape, and, once again, the air smelled of lemon.

Tortured by the thought of leaving Fredericka just when I had a chance with her, I left the yacht to look for a room in Perama. I quickly found a room with a view of the harbor and the anchored *Hydra*, over the top of a maze of whitewashed stone walls. The next morning she was gone. After another butter-and-honey breakfast, I boarded the first bus to anywhere. It was headed for Methymna, a larger town to the north.

From the window of my rented room in Methymna, I could see the mainland of Turkey, possibly an area close to where Agamemnon's ships anchored 2,500 years ago during the Trojan War. My landlord, Nikos, was a sheep farmer who loved to take me to one of the local cafés to discuss Greek history, politics, and philosophy, three subjects of which I knew close to nothing.

He was a peculiar man. His roughly shaven face, wrinkled by excessive sun exposure, made him look much older than his true age, which might have been early forties. He was a man of strong beliefs, especially when he talked—as he often did—with a bottle of ouzo in his left hand. He was a street thinker, a man with no formal education, following an old lost tradition of early philosophy started by Pythagoras and Hippocrates on Lesbos's neighboring islands, Samos and Chios. He would order a bottle of ouzo, fill a shot glass, take a swig from the bottle, take a puff from a cigarette, and ask a multi-layered question involving mathematics, science, religion, or all three at once. A cigarette was always in the hand that held the bottle. To take a puff, he somehow managed a trick of never letting go of the bottle. Unstable ash stems would grow till they fell to a tin ashtray from their own weight. His right hand never left the shot glass. Within minutes of his entering the café, his loud and rough smoker's voice would attract nearby fans and friends to listen while standing. He had an able philosopher's talent for asking those deep questions that had no plain answers.

One day he invited me to the café to show me something he had dug up when he was a boy. It was a golden ring with cryptic engravings. There was a story behind this ring that was well known among the villagers. Local legend had it that Julius Caesar came to Methymna just after having had a homosexual relationship with King Nicomedes and that the ring belonged to Caesar.

Nikos showed me the ring. It certainly did seem ancient, with gold that was sea-beaten for a thousand years. But one could still make out the engraving of a collection of symmetric polygonal lines, hinting something curiously mathematical.

When Nikos learned that I was studying mathematics, he excitedly relayed a story about another mathematician who had visited the village the previous year and who had given a mathematical explanation. It seemed that the ring was a commemorative engraving of an idea for calculating π, the ratio of the circumference to the diameter of a circle. There were three concentric polygons neatly inscribed in a circle: a triangle, a hexagon, and a dodecagon. Presumably, the next polygon would have twice as many sides, so the nth polygon would have $3 \cdot 2^n$ sides. The side of any $3 \cdot 2^n$–sided polygon can be computed by simply using the Pythagorean theorem.

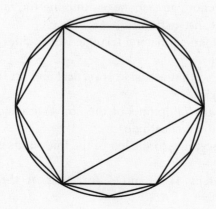

What did the ring really signify? Did it suggest that if one were to increase the number of sides of the polygon indefinitely that the length of the polygon would approach the length of the circle? That might seem like the logical conclusion to us twenty-first-century thinkers, but perhaps the thought was simply to approximate the circumference of the circle by a polygon of a large finite number of sides.

I was amazed to learn that Nikos had thought about this before. There are two ways to view what is going on. One is to simply assume that the radius of each polygon is a unit (the same as the radius of the circle), pick a polygon with a large number of sides, calculate the circumference of the polygon, and let that be the approximation to π.[1] The modern way—modern, that is, in the sense of post-seventeenth century—is to calculate the circumference of an n-sided polynomial, where n is nonspecific, and watch what happens when n increases toward infinity.

Nikos argued that just because the sides of the polygon get closer to the circle as n increases, it's not at all clear that the total length of the polygon gets closer to the circumference of the circle. How's that? He had a point. Remember the geometric argument in Chapter 4, showing that $1 = 2$? We took an equilateral triangle with sides equal to one unit, then constructed two equilateral triangles with sides half as long, and then constructed four equilateral triangles with sides half as long again (see illustration on page 68). We continued the construction, at each stage doubling the number of triangles. Here are the instructions:

1. Start with an equilateral triangle whose sides are one unit in length.
2. Orient it so that one side is vertical and on the right side of the triangle.
3. Connect the midpoints of the nonvertical sides to the midpoint of the vertical side.
4. Erase everything to the left of the midpoints of the nonvertical sides.
5. Repeat steps 1, 3, and 4 for each of the two triangles constructed.

6. Repeat steps 1, 3, and 4 for each of the four triangles constructed.

7. Repeat steps 1, 3, and 4 for each of the eight triangles constructed.

8. Repeat steps 1, 3, and 4 for each of the previously constructed triangles.

Notice that as the triangles are shrinking, the total length of all the nonvertical sides is always equal to two units, while the length of the vertical side is 1. Also notice that the nonvertical sides form a jagged (sawlike) line that is approaching the vertical line. If we carry out this recipe forever, the jagged line becomes the vertical line, and, therefore, 2 must equal 1.

Each new figure in the construction was narrower than the previous one, so their widths approached 0. This suggested that the sawtooth side would eventually meet the straight-line side as the number of teeth increased and, therefore, that $1 = 2$.[2]

What could be the difference between letting the nonspecific n in the n-sided polygon increase and letting the nonspecific n in the saw tooth increase? Every mathematician will agree that the first case leads to a valid approximation of π, but nobody will ever agree that $1 = 2$.

Nikos claimed that he would not sleep at night until I resolved this strange paradox for him. But I could not. I was as surprised as he. In the end, it was I who could not sleep. I lay awake most of the night thinking about the problem. What could be the difference between the two constructions? The only difference I could come up with was that, in each new subdivision of the saw-tooth construction, the total length did not change, whereas, in each new subdivision of the polygon construction, the total length increased. Was that the answer? Surely not. The lengths in the pictures behind both constructions seemed to converge. And yet mathematicians since Archimedes believed that the successively increasing lengths of polygons inscribed in a circle of radius r approached $2\pi r$.

The puzzle and its mystery provoked me to learn more about *analysis*, that branch of mathematics that deals with infinite processes. The answer was a great surprise, the kind that nags at the core of normal intuition so much that it releases just enough doubt about the basics of the space we live in to make one feel lost in a wonderland where saw teeth could be sandwiched between two walls of forever narrowing distance with no change in the total length of teeth.

CHAPTER 8

Zindo the Trojan Superman

Zeno's Paradoxes of Motion

He thought he saw a paradox
*   With time and space entwined:*
He looked again and found it was
*   A teasing of the mind:*
*   "And how could Achilles win," he asked,*
"By starting from behind?"
—J.M.

The struggling old Mercedes bus from Mytilene could not navigate the steep hills and sharp turns of narrow Methymna streets, so each afternoon at about 3:00, it stopped just at the edge of town. I took to the habit of greeting it with a tall, attractive woman from the tourist bureau, as if I were expecting someone. I wasn't. I was drawn by a native thirst for cultural companionship, a fantasy of a mysterious surprise arrival. That fantasy magically materialized one day when a short, brawny man bounced off the bus swinging an overstuffed orange rucksack to his back.

The coincidence could not have been more surprising. Jeanot was my good friend and classmate in Paris who had an exceptionally gifted talent for mathematics. I often relied on him for hints on class assignments. There was a moment of disbelief in seeing him. He had grown a scraggly red beard that made me hesitate a greeting, but he immediately recognized me as if he expected me to be waiting for his arrival.

"Hey! Jojo," he called. My Paris friends called me Jojo.

"Jeanot?" I hesitantly responded. "Is that you?"

Contradicting fictitious stereotypes, this mathematician was wildly adventurous. That evening he concocted a plan to rent a caique and cross over the narrow stretch of channel north of Methymna to the Turkish coast.

"Let's climb Mount Ida," he said breathlessly, after wildly knocking on my door late at night.

"We'd be entering Turkey illegally," I warned. But he countered my objection with a who's-to-know remark and continued his plan with the exuberant energy of an astronaut about to be launched into space.

Early the next morning, he convinced a fisherman to take us across the channel in a splendid caique that was geometrically painted in blues, pale yellows, and black; the head of a brightly colored, bulging-eyed bird mounted to its bow gave us courage to cross the Muslim channel to the rugged coast of Turkey.

We left Methymna with enough food and water for two days. The fisherman was to pick us up two days later at 4:00 in the afternoon at a place on the extremely rocky shore where he left us. We waded to shore barefoot, over the barnacled sharp rocks and broken cockle shells, as small waves drenched us to the waist, and then climbed the rocky hill to find a broken macadam road. Walking to the nearest village, we were taking in the beauty of the surrounding landscape—to our right were hills of bushes resembling laurels. I was watching a far-off white-sailed yacht interrupting the continuous blue of the Aegean when a farmer guiding a horse-driven hay wagon stopped to give us a lift to a tiny village by the sea. The gentle heat of the sun felt soothing after our cold and choppy channel crossing. Snuggly sinking into the comfort of a dry mound of hay, facing a cobalt blue sky above a dreamlike landscape high above the sea, mesmerized by the sound of horseshoes clopping against the macadam surface, I thought about the paradox, like a nerd thinking about math at a dance.

"I cannot understand," I said, turning to Jeanot after describing the suspected paradox. "What is the difference between the trick saw-tooth argument that $1 = 2$ and the acceptable approximation of π by the exhausting construction of an infinite-sided polygon?"

Jeanot fell asleep without answering. Did he fall asleep because the problem was so embarrassingly simple? Wasn't calculus the answer? Weren't we taught that functions approach limits without ever becoming their limits and that, therefore, the limit of inscribed n-sided polygons could approach the circle without ever actually becoming the circle? I felt that that was the answer, and yet, as I thought more about it, that answer seemed to devilishly slip away into that foggy, ungraspable place beyond conceptually finite margins.

I was still thinking about that elusive problem while watching the yacht come closer to shore when the hay wagon hit a particularly nasty crack in the road. The jolt suddenly woke Jeanot, who jumped up from his sleep with an answer.

"The closed polygons only approach the circle," he started, but then paused for a few seconds, as if he needed the language to catch up with his thinking. "They never become the circle," he said. "The same with the edge of your saw. It will never actually be the straight line; it will only approach it. So your paradox is this: Though your saw-edge constructions approach the straight line, the *lengths* of your constructions do not approach the length of the line. Is that your paradox?"

"Yes," I agreed, realizing that he clarified my own question. "But the lengths of the closed polygons inscribed in a circle approach the circumference of the circle! What's the difference between the two cases?"

"Look," he began with some visible annoyance, "it's clear that no saw edge, no matter how many teeth it has, will ever be the straight line. After all, even after an infinite number of successive constructions, the only points common to both the saw edge and the straight line are those that are distances that are multiples of $1/2^n$ units from the bottom, for any whole number n."[1]

"Yeah, I agree," I said. "But that is always the case when one takes a limit. In calculus, things approach without ever reaching."

"Well, the same is true about the infinite-sided polygon, whatever that is. The only points common to both the circle and the polygon are those that are distances $\pi/(3\times2^n)$ units from the top of the circle, where n is any whole number."[2]

"Yes, yes, I understand all that, but why does the arithmetic of the construction converge in one case and not in the other?" I reverently asked. "Why should the saw edges have the appearance of geometrically approaching the line while their numerical attributes[3] stay far from those of the line?"

"Good question." That remark made me feel a bit better. "The geometry of becoming is merely appearance. We don't really *see* what is happening beyond the misty edge of the finite world."

Until that moment, I thought I understood calculus. But at that moment, my understanding seemed to slip away. One minute, I understood it; the next, I felt fully ignorant.

It was quite some time before I was able to understand the difference between the appearance of convergence and the numerical mechanics of convergence. In geometry, there is no sharp edge to the finite world; there is a nebulous region that is beyond our visual— even conceptual—capabilities; we commonly rely on mental constructions of symbolic guides to anticipate the behavior of numerical attributes. Just because the developing saw edges appear to be geometrically becoming the line, it does not follow that their lengths are numerically approaching the length of the line; in fact, the length of the developing saw edges might wander far from the length of the line.

There is no fringe, no border, no edge, no margin where the finite world meets the infinite. There is no specific moment when a finite construction becomes the infinite construction, just as there is no specific moment when the saw edge becomes a straight line. It never becomes a straight line; it is always becoming.

That still leaves the question of why the numerical convergence is valid in one case and not in the other. Why do the lengths of the progressive constructions of inscribed polygons approach π while the lengths of the progressive constructions of saw edges approach a number other than 1? The simplest answer is this: In the polygon case, the process is controlled by the successive computations of the total lengths, which clearly approach a number as the number of sides approaches infinity. However, in the saw-tooth case, we expected the total length to converge because we assumed that if the (two-dimensional) space sandwiching the teeth shrinks, the total (one-dimensional) length of the teeth will also shrink. The surprise is this: To know that a length converges, it is not enough to simply know that the space it's confined in is shrinking. After all, we could fit an infinitely long line in a finite space.

We came to the edge of a hill overlooking the sea and Babakale, a drab village rambling downhill to a delightful cove of colorful fishing boats. Then we turned off the macadam to a dusty dirt road leading to the village center, passing unattractive houses with red tiled roofs. The farmer stopped the draft horse and let us off at a café decked with rows of rusty tin cans filled with geraniums, chili peppers, and basil. "Babakale," he announced, repeating the name several times to make sure we understood.

It was that special day of the month when a traveling barber-dentist set up his chair at the café to shave the men who needed shaving and pull the teeth that needed pulling while a young man with a splendidly trained spider monkey entertained the children outside. An ageless man in a graying thermal undershirt, sitting outside smoking a hookah, rose to offer a drink. It was customary in those days for Greeks and Turks in small villages to buy drinks for strangers—and sometimes even to offer elaborate meals. We neither spoke nor understood Turkish, so conversations usually involved a mixture of acutely broken English, French, and unintelligible words that nobody could identify.

"You know that this is a very important town?" said the man in French as he put down his hookah. This was his greeting. It was as if we were already in the middle of a conversation about historical geography. No greeting, no expression of wondering about two strangers being dropped off a hay wagon at the center of his little village.

"A very beautiful town," I enthusiastically replied out of politeness, not knowing what my host was referring to.

"This is the birthplace of Zindo. You know who he was?"

I thought he said Zeno, so I answered, "Sure, he was a mathe-matician—err, philosopher," correcting myself, thinking there was a difference.

"No, no," he retorted. "He was the inventor of football! He made this town great!"

Before long, a typical crowd of village bystanders grew. A few men moved their hookahs and chairs outside the café to come closer. My host continued to relay myths about the great man. Others in the party would make comments in Turkish and wait for our host to translate them. Some myths were so preposterous that we felt the group was either mocking us or stoned. There was the one about Zindo organizing a game of football between the Turks and the British in Gallipoli during a break in the fighting for the Dardanelles in World War I. Someone claimed that that was how the British were first introduced to football.

"Zindo was a *Truv* (Trojan)," the ageless man said, looking sincere. Some villagers claimed to descend from Trojans, just as Lesbians across the channel prided themselves as being descendants of Agamemnon. I volunteered to say that there was another great man, a great philosopher, who lived sometime back in the fifth century B.C. Taking liberty to exaggerate, I said that he might have visited Babakale.

We spent the night in a small room above the café. It was a cheerful room, yellow, with crooked sandstone walls and a fabulous view of the magnificent sunset over the sea through a solitary window facing west above the dirt main street. After dark, we could see caiques moving out to sea in circles, with lanterns brightly beaming back to shore like fireflies on an August night. But in the morning, when the fishermen returned and the cocks crowed loudly, when donkeys loaded with sacks of lemons complained and the sun promised another day, I rose from my bed for fresh air from the window. Throwing open the shutters, I could see the white-sailed yacht. It had come closer to shore. On the street below, amid a line of donkeys, motorbikes, and group of curious children, was a London taxicab.

"Jeanot," I called, "wake up! Look what's on the street."

Jeanot rubbed his eyes and came to the window to see the strange vehicle below.

"Do you suppose that someone actually took a cab from London to Babakale?" I asked. Somewhere in our friendship, my respect for Jeanot's mathematical wisdom evolved to a point that I thought he could tell me the answer to anything.

"No," he answered, "of course not. Surely some local important person bought the car in London."

We dressed to find out what sort of breakfast we could get at the café. A young Englishman sat outside, surrounded by the same group of hookah smokers of the night before. Simon Parks was dressed in khakis with a white V-neck pullover, the kind one sees at cricket matches. He was elegantly drinking his coffee with his pinky extended when Jeanot and I came down.

"You must be English," I said in a greeting tone as we approached.

"Yes," he said with a smile. "Who else would be wearing a white sweater driving a London taxicab in Turkey?"

"A Turk, I suppose."

"Simon Parks," he volunteered, as he laughed and rose to shake my hand. We joined him for some stale bread, butter, and coffee.

"What's with the taxicab?" Jeanot ventured.

"I drive rich tourists to these interesting coastal towns off the beaten tourist paths. The cruise ships stop in Kumkale harbor for the day so tourists can see the excavations at Troy. They book tickets through a bloke in London, and I take them on a tour through notable villages like Babakale in luxury. The taxi rides on air suspension, has air conditioning and fresh flowers, is packed with English food, and has a small bar. Come take a look."

Simon proudly walked us to the taxi and opened a rear door. The Queen's car could not have been more luxurious. Velvet red curtains draped its windows; two small bracketed vases, white roses and ferns in each, projected from its door frames; and a chrome bar unfolded from the low wall separating the driver. But from outside, it was indistinguishable from a London taxi.

"Where are the tourists?" I asked, doubting my thoughts of the possibility that they could be asleep in ordinary shabby rooms above the café.

"It's my day off," he said. "And on my day off, I explore to find the best towns to visit when the tourists are on board. This town is interesting for its contradictory characteristics. It's both unattractive and beautiful. It all depends on which way you look. Stand on the hill and look out to the sea, and you think that you have never seen anything as beautiful; stand in the cove and look at the boats, and you feel the same way; but stand in the cove and look up at the town, and you wonder how such a beautiful piece of landscape could have been made so unattractive. You know, this is the most westerly point of Turkey. Come later in the day," he continued with a giggle, "and you can join the old folks for intelligent conversation over smokes of the hookah."

"We had plenty of intelligent conversation last night, thank you very much," Jeanot said with an acknowledging giggle.

"Well then," Simon said, "would you like to see the best towns and villages?"

The taxi moved over the dirt roads and broken macadam like a Rolls. We had a few kurus, but whenever we bought food, some Turk would offer to pay. And if the language were compatible, we would discuss the local homebrewed philosophy for hours. But it was Simon who seemed to know the local history better than any of the villagers. We stopped in Behramkale, a coastal village near the ancient site of the Assos acropolis, with its Temple of Athena and terraces of ruins descending to a little idyllic harbor. Here the golden cliffs reflecting the afternoon sun made the water seem green, but we could see the whole island of Lesbos and imagine her a beautiful mermaid peacefully bathing on her back.

Simon pitched a spiel as if we were his tour group. We believed him.

"In the sixth century B.C.," he began, "the Ionian coast had several great shipping ports, its coastal towns of great travelers, explorers, tradesmen, and even pirates, who went to the ends of the known world and returned with news, stories, ideas, and musings. Ionians traveled east as far as Phoenicia (roughly present-day Lebanon and Syria), south to Egypt, and west to Spain. Understandably, it was a time when Ionians—and Greeks, in general—were interested in the beliefs and customs of other nations."

But then, at another café in Behramkale, surrounded by another group of men smoking hookahs, once again, unexpectedly, Zindo's name was mentioned—only this time, one of the Turks spoke reasonable English.

"You are from England?" an old man asked, looking at Simon after taking a puff.

"Yes, I am," Simon said.

"Well, then," he continued, "you must have heard of our great football player, Zindo." The old man claimed to have learned of Zindo during the Gallipoli campaign. "I was there playing football with the British." He told incredible stories of how the British and Turkish trenches were so close that the enemies could call to each other. Every Sunday, the fighting would stop for an exchange of cigarettes and a game of football. On Monday mornings, the fighting would resume.

When I told Simon of the conversation we had on the previous night in Babakale, he smiled and said that that seemed to be the local folklore. He had heard the tale before.

"There is a local ghost legend," he said. "Sometime back in the early days of the Ottoman Empire, a sailor by that name earned the reputation of hero for some heroic deed. As a result, he is reincarnated in every generation for new legends of heroism. His tomb is somewhere near Babakale. Whenever Turkish fishermen pass his tomb, they throw breadcrumbs into the sea. Don't ask me why. That's just the way the story goes."

"He was the inventor of football," said the old man, taking another puff.

"Yes, yes," Jeanot began in a feeble attempt to argue with the man. "But there was another Zindo, a philosopher who lived in the fifth century B.C." The old man translated the argument as Jeanot continued, though the other stoned street philosophers didn't seem to be listening.

"Motion is impossible, Zindo concluded," said Jeanot. "Zindo and a turtle were about to have a race." Several men started coughing and laughing. "The turtle argued that he should be given a head start because it was well known that Zindo was fast. 'But,' said Zindo, 'if I give you a head start, I will never be able to catch up with you.' 'How is that?' asked the turtle. 'Whenever I get to the place where you were, you will no longer be there. You will have moved further ahead.'"

The men looked puzzled. Simultaneously, they picked up their pipes and resumed smoking.

"Now Zindo would not be so famous if that were his only paradox," Jeanot continued. "He had others, plenty of others. He could show that you cannot get from this side of the road to the other."

The men put down their pipes and looked across the road while Jeanot stood up and went to the edge of the road to demonstrate.

"You see," Jeanot continued, as if someone asked him to continue, "to get to the other side of the road, you would first have to get halfway across the road." Jeanot took a few steps to the center of the road. "But you would then have another half a road to cross," he continued. "So you try to cross that remaining half. But to do so, you have to get to a point halfway across the remaining half." He took a few more steps and realized that he had already lost his audience to competing illusions.

By the time he took the next step in his illustration of the infinite, it was clear that nobody but the translator, Simon, and I were listening. The smokers made themselves busy filling their pipes with what we could only assume to be tobacco and water. Jeanot went on, knowing that he had lost most of his audience.

"I'll give you another one," he said, picking up a stone from the ground and looking only at Simon. "Zindo said that if I throw this stone as hard as I try, it will never leave my hand." He pretended to throw the stone.

"You see, Zindo argued that if it ever did leave my hand, he could freeze an instant in time to take a look at it—as if you had taken a photograph of it—and notice that, at that instant, it would not be moving. That led him to conclude that it couldn't go anywhere. After all, he argued, whenever you look at it, it is stationary."

Our English-speaking Turk didn't catch much of Jeanot's speech, but he caught enough to retort.

"Maybe *you* can't throw a rock across the road," he said, "but Zindo, our football star, sure could." And with that, he said something to a small boy who was standing nearby. The boy left and returned after several minutes to hand the Turk a stack of magazines.

"Proof!" said the Turk, hitting the top magazine several times with his index finger. "It's right here. Ah, but you don't read Turkish."

He held up the top magazine to show the title: "Zindo in Monaco." There was a picture of a dark man in a paisley bandana on a motor scooter coming fast around a hairpin turn on a cliff road high above the sea. A young girl on the back seat held his muscular arms, her hair trailing in the wind. It could have been an ad for the motor scooter, which was clearly marked *Vespa*, but it was not. The magazine was filled with black-and-white pictures of the same man, and his name was clearly Zindo. It was a photo-romance magazine that looked like an American comic book with photos in place of cartoon illustrations. Zindo was a photo-romance hero, a Turkish superhero. He was the inventor of everything. One issue showed him as a cowboy in pursuit of a villain. In one frame, he was riding a sweating stallion fast against a torturous wind. In the next, he was side by side with the villain, leaning over to grab him by the throat.

The real Zeno was a relatively tall and handsome man, a bit older than Socrates and about twenty-five years younger than his lover, Parmenides. He was an intellectual revolutionary who read frightening mathematical treatises at the market place and occasionally at the Great Panathenaea festival. His paradoxes of motion were striking and devastating. Socrates himself would come to hear him read.[4] Perhaps he did not mean to cause harm but simply wanted to stir philosophers to address the problem of infinity. After all, there was an ancient tradition to question for the sole purpose of stirring further thought. It seems plausible that Zeno had a similarly lofty purpose for his so-called paradoxes.

His classic arguments are against motion. Jeanot argued that a man cannot cross a road before reaching halfway across, and cannot reach halfway across before reaching a quarter way across, and so on; therefore, not only can the man not cross the road, but he can't move at all. One can't help wondering about those strange delays at those arbitrarily chosen milestones across the road. Surely, there is no stopping at those milestones. In some sense, Zeno has tricked us into imagining an infinite number of delays in moving. Why should we be hoodwinked into thinking that objects move by counting? Even stranger, Zeno is suggesting that we can freeze instants of time and still talk about kinematics. The paradox should then focus on the question of how an object can pass through infinite decelerations and accelerations.

Another Zeno argument was designed to demonstrate that an arrow could not move because every instant you looked at it, it is stationary. In other words, if you fix time, you must fix a position in space. One might argue that the very notion of fixing a point in time is absurd and that it makes no sense to say "An arrow appears stationary at any point in time." But in mathematics, time is simply a variable that can be fixed by simply declaring it to be some number of units of time from some starting time. We have mathematical formulas that tell us where the arrow is at any time t, so if we let t equal some specific time—say, two seconds after leaving the bow—we should know the exact spot where the arrow is when $t = 2$. But is there any such thing as two seconds? We know that if we really try to take a picture of the arrow when $t = 2$, we must have the shutter open for an entire interval of time surrounding $t = 2$. The shutter is a physical object that cannot open and close at the same instant. In other words, no matter how fast the shutter can open and close, the picture will be blurred in the direction of the movement of the arrow.

But this makes things even more puzzling, for it means that mathematics is not representing reality! We can stop time in a mathematical formula by evaluating the formula when the time variable equals a constant, but we cannot stop time in the physics of the real world. It means that the mathematical representations of physics are simply mathematical models. So, it would appear that the key to understanding Zeno's arguments is to understand the connection between what it means, both mathematically and physically, to let the time variable be equal to a constant.

David Hilbert and Paul Bernays put forward their solution of the Zeno paradox:

> Actually there is also a much more radical solution of the paradox. This consists in the consideration that we are by no means obliged to believe that the mathematical space-time representation of motion is physically significant for arbitrarily small space and time intervals; but rather have every basis to suppose that mathematical model extrapolates the facts of a certain realm of experience, namely the motions within the orders of magnitude hitherto accessible to our observation, in the sense of a simple concept construction, similarly to the way the mechanics of continua completes an extrapolation in which a continuous filling of the space with matter is assumed…. The situation is similar in all cases where one believes it possible to exhibit directly an [actual] infinity as given through experience for perception…. Closer examination then shows that an infinity is actually not given to us at all, but is first interpolated or extrapolated through an intellectual process.[5]

We were not permitted to pay for drinks at the café and could no longer—as we had done until then—politely refuse our host's insistent invitations of puffs from his hookah. So, as the day moved on into the late afternoon, when, once again, the orange sun seemed buoyed by its own dazzling reflections on the sea, under hairline strands of faint clouds shimmering gold and reds, we smoked water pipes with our new friends. We were told that the pipes contained tobacco but were not told what else. With sensations of ghostly weightlessness slowly rising through our bodies, we spoke words of a language we did not understand, though what we said seemed to make flawless sense. When I picked up *Zindo in Monaco* from the stack of photo-romances, I discovered that the cover was in color after all. Zindo's paisley bandana was bright red, as were the lips of his admiring companion.

Two blasts of a far-off ship's whistle brought my attention out to sea. Looking westward just beyond the small cove, I saw an impressive yacht with a distinctive black hull and white sails. The cove would have been too small for the large yacht's deep keel, so it anchored in choppy waters some distance from land, or so it seemed. Thinking she was the *Hydra*, I reeled down to shore to get a closer look at its formidable bowsprit. The visible silhouette of Lesbos, that beautiful mermaid bathing on her back, summoned thoughts of Fredericka's kiss, the one and only one I was permitted to have. Once again, I felt the dead octopus around my neck and the fantastic moment when the silent undulating underwater plants were the only witnesses to our embracing naked bodies. When I reached the cove, I could no longer see the yacht. It was as if she had never been there at all.

CHAPTER 9

Finding Pegasus

Do Irrational Numbers Exist?

He thought he saw the root of two
*   Expressed as rational:*
He looked again and found
*   It to be aberrational:*
"And all the time," he said, "I thought
*   It didn't exist at all."*
—J.M.

J eremy was only ten when he dived from his father's yacht off the coast of Pine Cay, a private island north of the Dominican Republic owned and managed by his father's exclusive Meridian Club. He snorkeled through the shallow warm, crystal-clear waters to take underwater videos of schools of yellow angelfish and spectacular coral, unaware of the blue shark furtively swimming behind him. A brave fisherman who happened to be nearby rescued him, but not in time to prevent severe injury. It took years for Jeremy to recover, but by the time he enrolled in my calculus class in the fall of 1985, only a slight limp was noticeable. He was a nervous, garrulous young man whose clothing style was in tune with his identity—a character in conflict with rebel, attention seeker, and geek. He wore a ten-gallon hat over his long, wild, curly red hair and always kept at least six pens in a pocket protector in his breast pocket. From his belt hung a leather case containing a harmless Texas Instruments pocket calculator.

Jeremy challenged almost everything I said in class but never questioned his calculator. If I said that the product of √23 (the square root of 23)[1] with itself is exactly 23, he would object. Whenever a calculation contained a radical number such as √23, he would quickly whip out his calculator like Tom Mix (who always drew his gun but seldom shot to kill), press a few buttons, and yell out the nine digits displayed by the machine as if some other answer would win favor if he were not quick enough on the draw. This calculator slinging was his typical classroom behavior. Privately, I tried to advise him to think that numbers have properties as well as measure. "Think of √2 as a number near 1.4 that has the wonderful property that its square is exactly 2," I would say. But it was no use. He could only see such numbers as being measurements that had to be tamed down to, at most, nine digits or, as he would in some cases accept, a simple fraction. Once, I gave him a problem that involved squaring √23. Instead of seeing that the answer was simply 23, he went for his trusty TI-85, entered 23, pressed the square root button to get 4.795831523, and then pressed the square button to get 23. "You see," he said proudly, "I got the same answer as you did by doing it on my calculator."

I couldn't let it go, so I asked, "What happens if you multiply 4.795831523 by itself? Would you expect to get 23?"

"Yes, of course! Isn't that what I just did?" He asked in a tone of surprise. He entered 4.795831523 and pressed the multiplication button, followed by 4.795831523 again. When he hit the equals button, the number 23 appeared. "You see, 4.795831523 times 4.795831523 equals 23!" he said with a puzzled grin.

"But, but…" I stammered, "how can that be? You just multiplied two numbers that end with a 3? Shouldn't you get a number that ends with a 9?" I asked, setting him up for my clincher.

"Maybe that's true when the number of digits is small, but you can see that it is not true when the number of digits is nine, as it is for 4.795831523."

At that point, I had to give up. Jeremy was too quick a calculator-slinger for me. He had grown up believing that his calculator was invincible. It was a tool whose answers needed no interpretation or explanation. It would have been too painful for him to hear that his calculator was cheating. I believe that he had no idea of how to multiply two numbers together without a calculator. The concept of infinity was so unacceptable to Jeremy that there was no way to convince him that π was not 22/7 (which he took to be 3.142857143)[2] and that 22/7 was an approximation that was established solely to simplify computations back in the days before calculators were invented.

Halfway through the term, Jeremy discovered a warm radiator to lean on from his desk in the last row. The leaves had fallen from the maple trees just outside the classroom window to reveal a spectacular westward view of receding hills vanishing into the sunset. Jeremy would often appear either to be gazing out the window or to be asleep, but whenever I was about to present a calculation, he would quickly wake, fully aware of what I had just said, and correct my calculation or make some objection. Occasionally, he would wake to say something the other students thought irrelevant. One day, during a lecture involving the Zeno paradox, Jeremy asked an alluring question. He asked, "At what point of a horse race does one horse become, inevitably, the winner?" He had just returned from Sienna, Italy, where he had observed the Palio, the colorful and dangerous horse race through three hills of steep, winding streets and alleyways of thirteenth-century Sienna, to a grand finish in a beautiful, expansive central square. His mind brilliantly tangled and twisted life with fable to implicitly link the Palio with Achilles and a tortoise.

Ten of the seventeen districts of Sienna compete on July 2 to commemorate the miracles of the Madonna Provenzano, and then again to honor the Assumption of the Virgin on August 16. It is an unusual horse race of jockeys riding bareback in colorful costumes. In the three days before the Palio, the horses are kept under guard and the jockeys are held incommunicado. It is a highly dangerous event, often involving injury and even death. If a horse is behind by one-half a length within one-quarter mile of the finish line, the lead horse will inevitably be the winner. Barring some catastrophe, such as a slip or fall of the lead horse, the trailing horse has no chance of winning because the two horses are likely to be moving at speeds that are very close to each other.

Jockeys have their racing strategies, and sometimes those strategies call for falling behind. The trick is to know the last second when it is still possible to catch up.

At first I puzzled over Jeremy's question, but I soon came to understand that it was mathematically enticing. I made some quick calculations about a hypothetical situation and came up with the following example: If two horses are in a race and one falls behind the leader by one horse length (say, ten feet) it is very difficult for the one behind to catch up with the other by the end of the race. Thirty-five miles per hour is a very fast speed for a two-mile race. If one horse is running at 36 miles per hour and the leader is running at 35 miles per hour, then an easy calculation tells us that it would take 360 feet for the trailing horse to catch up with the leader. You can see that it would be impossible to make up that distance in the last sixteenth of a mile (330 feet) of the race.

"Aha! It's the answer to the Achilles and tortoise paradox," I finally said.

Zeno of Elea, as we have seen, argued that even the "swift-footed" Achilles could not win a race with a tortoise if the tortoise were given a head start.[3] Suppose that the tortoise is given a head start of one mile. For Achilles to catch up with the tortoise, he must get to the one-mile mark. By then, the tortoise will have moved to a new position past the one-mile mark. As Achilles passes the mark, he must get to where the tortoise is, but by the time he gets there, the tortoise will have moved farther along. So you see that the process of overtaking the tortoise is an infinite succession of getting to the places where it once was.

"This is truly puzzling," I said. "Let's suppose that Achilles's speed is ten miles per hour, the tortoise's speed is one mile per hour, and the tortoise is given a head start of nine miles. Then, when Achilles runs nine miles, the tortoise is 9.9 miles from the starting line. When Achilles is 9.9 miles from the starting line, the tortoise is 9.99 miles from the starting line. At Achilles's nth attempt to catch up, the tortoise will be 9.99...9 miles from the starting point, a number having n nines after the decimal point. If the race continues indefinitely, the tortoise will be 9.99... miles from the starting point. The dots at the end of this last expression indicate that there is an infinite string of nines after the decimal point."

"What? This last expression has infinitely many digits! What could it possibly mean to add infinitely many digits?" Jeremy asked.

"Jeremy," I continued, "suppose that your number 9.99...9 had n nines after the decimal point. Can you imagine that?"

"What's n?" he asked.

"It's arbitrary. Imagine it to be representing any number whatsoever," I said with a feeling of falling down a rabbit hole to Wonderland. "n is n; it's any number, but you can't think of it as being fixed at any number."

I quickly realized that there was a deep concept here that I was not explaining very well.

"Okay, okay," he said with a grimace, letting me continue.

"Now, let n increase and watch what happens to the number as n increases," I said, happy to get over the last hurdle. "If it approaches some number—say, 10—it seems reasonable to say that the infinite expression is really representing the number 10. The more nines there are, the closer 9.99...9 is to 10, so we say that 9.99... $= 10$."

Clearly, Jeremy was not happy with the idea that the infinite decimal 9.99... is really 10. But he granted me the favor of accepting the fact by what he called "a wildly imagined infinite growing process." Unsure of whether I had educated him or he was merely indulging me, I gave him another way to look at the problem of the race between Achilles and the tortoise.

"Another way to look at the problem is to notice that Achilles travels $10t$ miles from the starting line in t hours and that $9 + t$ is the distance that the tortoise travels from the starting line in t hours...."

"Okay," Jeremy interrupted. "Achilles travels $10t$ miles from the starting point because he travels ten miles per hour, and so in t hours, he will have traveled $10t$ miles."

"That's right," I said.

"And the tortoise will have traveled $9 + t$ miles because he had a head start of nine miles and was traveling at only one mile per hour."

"Yes," I continued. "At the moment Achilles catches up with the tortoise, we know that the distance from the starting line for Achilles must be the same as the distance from the starting line for the tortoise, so $10t = 9 + t$. Hence, Achilles will catch up with the tortoise when $t = 1$ hour. And the distance traveled from the start by Achilles will, therefore, be ten miles."

This is the usual answer that mathematicians give when confronted with the Achilles and tortoise paradox. Does it resolve the paradox? There is still the question of how Achilles will be able to perform an infinite number of things in a finite amount of time. The difficulty is that we are confusing motion in space with the movement of time.

Jeremy suggested a new paradox. "Surely," he argued, "just by standing still for one minute, Achilles would be performing an infinite number of things in finite time; examine him at any one of the real numbers between 0 and 1, and he would be in the same spot. Does this mean that he cannot stay still for one minute?" Jeremy's argument was reasonable. He had deep and clever arguments about space and time but could not accept any abstract notions of number or infinity. It was not simply a matter of obstinate resistance to thinking differently or any obstruction to understanding the concept of infinity, but more a virtue of feeling right about something and sticking to his belief.

Jeremy is right. Mathematics does not exactly represent reality. Mathematical representations of physics are merely mathematical models that fit physics tightly in some places and loosely in others. Models are just models; they are merely helpful representations that can be mathematically manipulated to tell us something that might otherwise be too complicated and difficult to unravel. The Achilles and tortoise race unravels the answer to Jeremy's question: Why is it so hard for one racer to catch up with another?

Jeremy would politely argue that infinity is to blame. "It's not a paradox," he would say when confronted with the strangely confusing suggestion that a person cannot move from one point to another in a finite amount of time because he or she would have to "stop" at an infinite number of places before getting anywhere. "These are the things that happen when you let infinity argue for you. You cannot believe in impossible things." When it came to studying the fundamental concept of calculus, *the limit*, he politely refused to accept the idea that the rational numbers[4] were any different from the irrational. For Jeremy, it was simple: There were no such things as real numbers that were not rational.

I had great sympathy for Jeremy's rebellion against the infinite, the infinitesimal, and even the abstract. I came to realize that there was a legitimate obstruction to his acceptance of limits and infinite constructions. The shark attack had left him with impaired confidence that affected his schoolwork during adolescence and necessitated the help of a tutor. Jeremy frequently talked about his tutor, Ike, with genuine admiration. His father hired Ike to home-school Jeremy during his last two years of high school. Jeremy once told me that Ike had advised him to rely on calculators because it was impossible for them to make mistakes, but I had the impression that Ike had realized Jeremy's need for the certainty of a calculator to overcome his profound underlying anxiety.

Could I disguise infinite decimal expansions in a way that Jeremy could conceptually accept? I thought about my earliest confrontation with the idea of the infinite and realized that the concept had developed alongside an understanding of what it could possibly mean for a process to never end. Jeremy would have rejected any never-ending process, so my only recourse was to couch the concept in a finite scheme of algebra. "How would he react to a proof that $\sqrt{2}$ cannot be represented by a finite number of decimals?" I thought.

One of the earliest mathematical proofs I understood was the wonderful Pythagorean proof[5] that $\sqrt{2}$ is irrational.[6] Roger had taught it to me somewhere on the way to the Orinoco River. He had to repeat it to me three times before I had the feeling that a grenade had exploded in my mind. For the first time in my life, I understood how definitions of words are connected—"rationalness" was suddenly linked to "square rootness"—in a way that made me know that the world might be complex but understandable through the linkage of meanings. Still, I was puzzled because I did not understand what the irrationality of √2 had to do with calculus, a subject that dealt with infinite processes.

I was amazed at how quickly Jeremy was able to understand the Pythagorean proof that √2 is irrational.[7]

I repeated it three times before he fully felt the impact. *Reductio ad absurdum* (proof by contradiction, or showing that embracing the alternative results in an absurdity) is a type of proof commonly used by mathematicians since Euclid's time. If you want to demonstrate that a statement S is true, begin with the assumption that S is false (assuming that "false" means "not true"), and arrive at a contradiction somewhere in your argument. According to mainstream mathematical convention, you will have shown that S is true[8]. Some might argue that you will have only shown that S is not *not* true, while others might argue that "S is not *not* true" is the same as "S is true." The Mad Hatter might butt in, "Not the same thing a bit! Why, you might as well say that, "No swan is not black" is the same as "All swans are black.""

The proof that √2 is irrational relies on the fact that the equation $p^2 = 2q^2$ has no solution where p and q are both whole numbers.[9] In classical Greece, √2 was considered an extremely strange thing. It was something that could not have been measured, even though, by the Pythagorean theorem, it represented the diagonal of a square whose sides equal one unit. To measure a distance, one must have a measuring unit; any unit will do. So, assume you have a measuring ruler marked by meters, centimeters, tenths of centimeters, and so on. Perhaps the distance is the diagonal of a square with sides one meter in length, a perfectly reasonable thing to measure. But what happens? The measurement is more than one meter and less than two meters. So we must look at the finer subdivisions to get a more accurate measurement. How much is it over one meter? More than four tenths and less than five tenths. Still, we don't have a precise measurement, so we look at hundredths of a meter (centimeters). How much over one and four tenths is it? More than one hundredth and less than two hundredths. And so on.

The Pythagorean geometers knew that if this process were performed forever, it would still not have a precise measure of the diagonal. Plato's *Theaetetus* is one source that attributes this discovery to the Pythagorean mathematician Theodorus, who appears in the dialogue but doesn't say much. His friend speaks on his behalf:

> THEAETETUS: Theodorus here was proving to us something about square roots, namely, that the sides [or roots] of squares representing three square feet and five square feet are not commensurable in length with the line representing one foot.... ..

So, we can conclude that the Pythagoreans knew that this "thing" that we call "square root of 2" is not a *commensurable number*, meaning that there is no fraction of a unit that could measure $\sqrt{2}$. If the process of measuring the diagonal of a unit square stopped at, say, one hundredth of a meter (a centimeter), the common measure would have been the centimeter; that is, both the length marked on the measuring stick and the diagonal would both be exactly 141 centimeters. But the process doesn't stop.

On the one hand, the diagonal of the square exists as a length; after all, we can see it—at least, we think we can—without fear of being mocked by the Florentine astronomer Francesco Sizzi. On the other hand, it is not measurable. Imagine the surprise among the early Greek mathematicians who discovered that there are simple things that occupy physical space that are not measurable.[10] What can science do if the simple diagonal of a square is not measurable? Doesn't size correspond to number? All this suggests that there is no such thing as "a number whose square is 2." If there is, what is it? Perhaps it doesn't exist. But denying the existence of something only makes talking about it meaningless. Hence, denying its existence is also meaningless. And, therefore, it exists. "Oh dear, it's so confusing," Alice would say.

What does it mean to say that something exists? Mathematics takes a liberal view: An object exists if its existence does not cause a contradiction. In Pythagoras's time, there was no such thing as "a number whose square is 2." In fact, even negative numbers and 0 were not considered numbers. Jeremy would have been able to converse comfortably with the Pythagoreans. The idea of simply admitting such a radical to the club of things we call *number* is a modern one. If such an admission does not upset mathematics by leading to contradictions, why not admit it?

In the summer of 1965, I was traveling with a friend from Athens to Trieste on the Orient Express when we received news that a circus train had derailed several miles ahead. Our train was delayed for several days in Belgrade. We had no visas to exit the train but refused to spend more days than planned in an overheated, overcrowded car. So we decided to sneak off the train, hitchhike to Zagreb, and reboard when it arrived at Zagreb a day or two later.

An unmarked canvas-backed Mercedes truck stopped to give us a lift. The driver was headed to Zagreb. The journey by truck over very bad roads would take more than a day. So, we camped for the night on the side of a hill and met a shepherd boy who was no more than twelve years old. Over a small campfire, he pointed out the constellations; the clarity of his descriptions and his knowledge of the constellations was impressive. Growing up in a big city, I always had trouble distinguishing one constellation from another.

When the shepherd learned that I was studying mathematics, he asked something so striking and deep that I didn't know how to respond.

"Is there any such thing as a perfect square?" he hesitantly asked. "My teacher says that there can never be a truly perfect right angle. Is that true?"

I thought I could change the subject by pointing to the sky and talking about the constellations. "That's Pegasus," I casually remarked, pointing to the constellation of the winged horse. But I quickly realized that my diversion pointed to an answer to his question. "Is there such a thing as a horse with wings? I mean, how can a horse have wings? Does Pegasus exist?" I asked.

"Pegasus is a mythological winged horse," I said. "It doesn't exist, except, of course," I continued in jest, "that it hovers above tens of thousands of Mobil stations all across America."

"He was born from the blood of Medusa," he said. "Pegasus dug a spring on top of Mount Helicon, where Apollo and the Muses lived. That spring is the source of fancy and poetic inspiration. Do you want me to tell you about the other constellations? There is Cygnus and there is Aquarius," he said, pointing to each side of Pegasus.

Normally, existence is limited to objects that take up space. But, like the "idea of Pegasus," $\sqrt{2}$ does not take up space. Do we include it in the world of existing things? This suggests that the world must contain many more "things" than just those that occupy space. But what about the "thing" we symbolize by $\sqrt{2}$? It doesn't take up space. We would like it to exist, so we *make* it exist. It is part of the constellations in the mathematical universe.

If you see a person out of focus, you must assume that the problem is with your eyes, not with the person. A person cannot be out of focus and blurry to everyone who sees him. In Woody Allen's film *Deconstructing Harry*, about a dysfunctional neurotic writer who can function only through art, Robin Williams plays Mel, a character out of focus. Mel is blurry, so everyone else needs glasses to see him clearly.

"Look at yourself," Mel's wife suggests. "You're soft!"

We are struck by the humor of the situation because we simply do not have any real-world notion of perceptible objects being out of focus. If an object is out of focus for everyone who sees it, then how can we make up our minds about whether it exists? Nothing in our world fits the description of an object the size of a person who is out of focus. If every person sees Mel out of focus, is he, himself, out of focus? He becomes blurrier with each passing day. What if everyone's sight of him is so extremely out of focus that he is not seen at all? Does that mean he isn't there?

Bertrand Russell used the following story as an example of a paradox. A journalist comes to a small village in Transylvania and asks the barber, "Whom do you shave?" The barber replies, "I shave all and only those in the village who do not shave themselves." Upon leaving the village, the journalist puzzled over the question of who shaves the barber. The barber shaves anyone in the village if and only if that person does not shave him- or herself. So the barber shaves him- or herself if and only if he or she does not.

One way out of this paradox is to say simply that such a barber does not exist. In mathematics, paradoxes cannot exist. So, a mathematician, upon hearing this story, would simply claim that the barber did not exist and, therefore, accept the story while denying the premise that the barber exists.

> "He's dreaming now," said Tweedledee: "and what do you think he's dreaming about?"
>
> Alice said, "Nobody can guess that."
>
> "Why, about *you*!" Tweedledee exclaimed, clapping his hands triumphantly. "And if he left off dreaming about you, where do you suppose you'd be?"
>
> "Where I am now, of course," said Alice.
>
> "Not you!" Tweedledee retorted contemptuously. "You'd be nowhere. Why, you're only a sort of thing in his dream!"[11]

On the other side of the looking glass, Alice has a hard time convincing Tweedledee and Tweedledum that she is, indeed, real and not simply a sort of thing in the snoring Red King's dream. Tweedledee and Tweedledum have Alice thinking:

> "If that there King was to wake," added Tweedledum, "you'd go out—bang!—just like a candle!"
>
> "I shouldn't!" Alice exclaimed indignantly. "Besides, if I'm only a sort of thing in his dream, what are you, I should like to know?"
>
> "Ditto," said Tweedledum.
>
> "Ditto, ditto!" cried Tweedledee.

Is $\sqrt{2}$ any less real than Alice? If the Red King wakes, does Alice "go out like a candle"? If no person—not even the Red King—ever measures the diagonal of a square, does that mean that $\sqrt{2}$ does not exist? Alice should have read Descartes's *Meditations* for a good rejoinder. The greatest existential question of all is that of self. *Cogito, ergo sum* was Descartes's reflective conclusion after his meditation experiments in doubting everything, including his own senses and body.

Am I not so bound up with a body and with senses that I cannot exist without them? But I have conceived myself that there is absolutely nothing in the world, no sky, no earth, no minds, no bodies. Does it now follow that I too do not exist? No: if I convince myself of something then I certainly existed…. So after considering everything very thoroughly, I must finally conclude that this proposition, *I am, I exist*, is necessarily true whenever it is put forward by me or conceived in my mind.[12]

By simply raising the question of whether *anything* in the world exists, he came to know that he himself must definitely exist.

So what about $\sqrt{2}$? Like a spirit, it comes in different disguises. It is "the number whose square is 2," or the length of the diagonal of a square of length 1; but it could also be a solution to the algebraic equation $x^2 - 2 = 0$.

Such equations normally model real events. For example, if you have a square with four equal sides of length 1 and let x denote the diagonal (the line joining two opposite corners of the square), then the Pythagorean theorem says that $x^2 = 1^2 + 1^2$. This is equivalent to $x^2 - 2 = 0$. We can now say that $\sqrt{2}$ is just a "thing" that is equivalent to the size of the diagonal of a unit square. As we have seen, we can know what it is only by getting as close to it as we want, using finer and finer subdivisions of a unit. We can say that $\sqrt{2}$ is approximately 1.414213562, but only approximately. To get it exactly, we would need to have the trail of digits to the right of 1 be infinitely long. What could that possibly mean? Surely, Jeremy would not accept the notion that a number could be represented by an infinitely long trail of digits. (I should mention one curious thing about the never-ending number that is represented by $\sqrt{2}$: Its digits have no repeating pattern. If they did, $\sqrt{2}$ would be a rational number, which is either a number whose digits have a repeating pattern or a number with a finite number of digits.)[13] The irrational number can be on the one hand an observable object—namely the diagonal of a square—and on the other, a mythical winged horse that manifests itself in the idea of a never-ending sequence of digits.[14]

But if $\sqrt{2}$ could be represented through an infinite process that had a simple pattern, perhaps we could see its infinite character through a finite process. That might enlighten Jeremy and convince him that $\sqrt{2}$ is more interesting than just another boring number on his faithful TI-85. I took a chance and wrote $\sqrt{2}$ as 1 plus a fraction of fractions.[15]

$$\sqrt{2} = 1 + \cfrac{1}{2 + \cfrac{1}{2 + \cfrac{1}{2 + 1 \cdot \cdot \cdot}}}$$

This gives a repeated pattern representation of $\sqrt{2}$.[16] This is useful in computing the continued fraction as $\sqrt{2}$, for that computation comes from noticing that if

$$x = 1 + \cfrac{1}{2 + \cfrac{1}{2 + \cfrac{1}{2 + \cfrac{1}{\cdot \cdot \cdot}}}}$$

then

$$x = 1 + \cfrac{1}{1 + x}$$

It also demonstrates that, when dealing with infinity, the old rule of logic that says the whole is the sum of the parts no longer applies.

Jeremy understood the derivation of this magnificent representation of $\sqrt{2}$. I thought his objection was connected to the patternless expansion of $\sqrt{2}$ that Jeremy was used to and that this new representation, through simple repeats of the number 2, would be more acceptable, just as $9.99\ldots = 10$ was more acceptable. I also thought that I had finally cleared all the obstructions to his accepting the infinite. But I hadn't. He still could not accept infinite constructions because of a different problem. The mental conception of performing an infinite number of steps was being tangled and twisted with real time.

"But you will never be able to get the value of x by this process because it will take forever to do it!" he argued.

And, in a certain sense, he was right.

After all, the shepherd boy's teacher was right: A square is less likely to exist in real life than a winged horse. Surely, outside of the mind, nobody has ever constructed a right angle, let alone a real square. So what does $\sqrt{2}$ represent, if we can never construct it? Jeremy said, "Yes, we can say that $\sqrt{2}$ is the solution to the equation $x^2 - 2 = 0$, but that is just a name for something that we can never finish constructing!"

"Okay," I said, thinking that I had another angle. "We could turn the argument around, so as to start with $\sqrt{2}$. Represent the number 2 as an infinite nest of square roots, namely:

$$2 = \sqrt{2 + \sqrt{2 + \sqrt{2 + \cdots}}}$$

To show that this is true, write

$$x = \sqrt{2 + \sqrt{2 + \sqrt{2 + \cdots}}}$$

Then x is the positive solution to the equation $x^2 = 2 + x$. This comes from squaring both sides of the previous equation to get $x^2 = 2 + \sqrt{2 + \sqrt{2 + \cdots}}$ and noticing that the nasty term on the right is just x again. You can see that the only positive solution to $x^2 = 2 + x$ is $x = 2$.

I didn't expect it, but of course, Jeremy argued that what I did was invalid.

"When you square x," he said, "it's not clear that you are simply removing the top square root sign; after all, you are squaring an infinite nest of numbers—how can you do that in a finite stroke of time?"

When I thought about this, I realized that, in a certain sense, he was right again.

The next day, he came to me with an even more brilliant retort.

"You might just as well say that 2, 3, 4, and so on can be written as continued fractions. Wouldn't that be silly?" Every whole number can be written as a continued fraction.[17]

$$2 = 1 + \cfrac{2}{1 + \cfrac{2}{1 + \cfrac{2}{1 + \cfrac{2}{1 + \cfrac{2}{\ddots}}}}}$$

$$3 = 1 + \cfrac{6}{1 + \cfrac{6}{1 + \cfrac{6}{1 + \cfrac{6}{1 + \cfrac{6}{\ddots}}}}}$$

$$4 = 1 + \cfrac{12}{1 + \cfrac{12}{1 + \cfrac{12}{1 + \cfrac{12}{1 + \cfrac{12}{\ddots}}}}}$$

For my last attempt, I pitched an argument that sounded something like this: Just consider how many days are in a year. There is no reason for anyone to suspect that the number of days in a year is a rational number. In fact, it might not be any one number at all and might depend on which year we are talking about. Earth spins on its axis while it orbits around the sun. The number of days in a year is simply the number of complete spins Earth makes by the time it completes its path around the sun. That number is somewhere in the neighborhood of 365.242198. We have to make an adjustment every four years; otherwise, our seasons would be off by almost a whole day. But that adjustment is not quite enough. Even if we could make a perfect adjustment, the next year would still be off by an imperceptible amount. Even pissing in a river can change the tides and, hence, change the number of days in a year.

Immeasurable details are one of the many things that make the world interesting, according to the French movie *Le Battement D'Ailes Du Papillon*, written and directed by Laurent Firode. Here's what one of its characters says:

> Listen to me. There's not a gesture, even the most insignificant, that can't change the world. That man there decided to lie to his mistress like he lied to his wife on the simple toss of a pebble. You see every detail, every gesture, as slight as it may be, reveals an infinity of truths and thus has an endless repercussion and grandiose effects. You only have to piss in the sea to make the ocean rise. Don't they say the beating of a butterfly's wings over the Atlantic can cause a hurricane in the Pacific?

After that, Jeremy was silent. I thought that did it. I was wrong. His retort was that if one looks too closely at things, one sees emptiness.

CHAPTER 10

Some Things Never End

The Logic of Mathematical Induction

He thought he saw Achilles
    Racing with a tortoise:
He looked again and found it was
    Time in rigor mortis:
"How could Achilles win," he said,
    "If time keeps stopping for us?"

He thought he saw Zeno
    With his silly paradox:
He looked again and found it was
    Space and time that mocks:
"Achilles could have won," he said
"If only he wore sox."
—J.M.

Perhaps time and space are simply fruits of imagination, fictions to help us understand our own existence. After all, there is a kind of indefiniteness that comes with infinity. Perhaps infinity is a deception that helps us cope with our indefatigable lust for explaining not only our own world, but also all possible worlds.

To a computer, time is not infinitely divisible. Its chips work to the beat of a clock, so it can compute by doing one bit every one-billionth of a second, but no faster. It views the race between Achilles and his friend, the tortoise, in a world where time moves by finite increments. A film of Achilles's race with the tortoise that moves at sixteen frames per second would still show Achilles winning.

We do not think of large numbers in the same way that we think of small numbers. Infinity could be thought of in many ways: a boundlessness, an infinite line of dots passing off into space until it evaporates into emptiness, something repeating forever, a finite circular maze with no way out, or a densely packed line with points so close that between any two there is always a third. The mind can easily represent these types, but it cannot see them on the mind's viewing screen. Yet something phenomenally fantastic does happen in the mind. The more one thinks about the ramifications of infinity using some sort of visual representation, the better the mind grasps what those representations actually represent. Intuition about infinity starts as vague impressions but eventually develops the mind's representative images into a deeper intellectual understanding. That same intuition also gives the impression of something real when, in fact, there is nothing real about it. As Simone Weil wrote in one of her notebooks, "Appearance possesses the fullness of reality, but as appearance only. As anything other than appearance, it constitutes error."[1]

We can imagine an infinite collection of numbers trailing off to a vanishing void, and logically talk of what happens when they reach their limit, indefinitely splitting intervals of space and time in the mind's eye, but eventually our images blur and repeat. Eventually, intervals between our numbers look so small that we cannot distinguish one number from another. So, at best, we imagine some minimal interval of finite but small size. But an infinite collection of finite intervals is infinite.[2]

One could say that all mathematics always takes place in the mind. It is a mental construct and can have no existence, except through mental activity. Symbols, equations, and theorems are merely the means by which mathematics is communicated. They form the sentences that add up to the story. Man has created the natural numbers as a mental device to record and explain the real world. But the natural numbers were constructed to permit the mind to operate, investigate, and agree with other minds. Once we started counting, there was no stopping. We were forced to accept a number system that is infinite.

Descartes said, "We can understand a triangle to be a figure made from three lines, but when we imagine it, we see before us in the mind's eye three lines with ends joined." Contrast this image with what we think of when we think of a chiliagon (a polygon with one thousand sides). We cannot see a thousand-sided figure before us in our mind's eye in the same way we see a triangle. So what do we do? We imagine something, but surely not a figure with one thousand sides. What we do is mentally construct a representation of a chiliagon, which is far from the picture of the real chiliagon. In fact, this representation is in no way different from what we construct when we want to think of the myriagon (a polygon of ten thousand sides) or a hectagon (a polygon of only one hundred sides). Hume puts it this way:

> When you tell me of the thousandth and ten thousandth part of a grain of sand, I have a distinct idea of these numbers and of their different proportions; but the images, which I form in my mind to represent the things themselves, are nothing different from each other, nor inferior to that image, by which I represent the grain of sand itself, which is suppos'd so vastly to exceed them . . . whatever we may imagine of the thing, the idea of a grain of sand is not distinguishable, nor separable into twenty, much less into a thousand, ten thousand, or an infinite number of different ideas.[3]

How can such representations be useful when they can't distinguish one figure from another with a difference of nine thousand sides? They are useful in cases when confusion between the chiliagon and the myriagon does not matter. In those cases, the mind simply creates a cognitive image that suggests a many-sided polygon. If and when a distinction is needed, the mind will create new cognitive images to distinguish between a one-thousand-sided polygon and a ten-thousand-sided polygon. It might be very hard to express what we see when we evoke the chiliagon into consciousness. But we do construct something. What? It might be a symbol, which might be formed simply by the word *chiliagon* or the image of a polygon with a large number of sides, or part of one. Or it might be simply a fuzzy picture of a squiggly line that has little to do with the real chiliagon. What we see is what Jung called a private symbol, fashioned from our personal experience of countless meanings and limitless emotions.

When we picture the square, we do better. I would guess that the private symbol of a square for most humans is a realistic cognitive image of a square. Moreover, it's better than that. The human's cognitive image of a square is the ideal square that does not exist outside the human mind. It is part of a constellation of personal cognitive experiences.

All human beings have the ability, which psychologists call subitizing, to immediately distinguish among one, two, or three items in a field of vision. We now know that four-day-old infants have that ability. All humans can accurately distinguish the number of objects in a collection that contains fewer than five objects. But most humans cannot quickly tell the difference between collections of fifteen and sixteen objects. So we carry with us in our heads this fuzzy representative image to use for both fifteen and sixteen, and not the images 15 and 16, which are simply good names for those images. Moreover, we should not think that this universal ability is simply a process of counting or evaluating, but, rather, more a sense of number itself. Some psychologists will even argue that this ability is innate.

We can imagine the triangle in the mind, but when we imagine a chiliagon, our mind vaguely pictures some kind of approximation—but we can clearly understand both the figure of the triangle and the figure of the chiliagon. *Imagining* is more demanding and difficult than *understanding*. Descartes dealt with this difference in this way: "When the mind understands, it in some way turns toward itself and inspects one of the ideas that is within it; but when it imagines, it turns toward the body and looks at something in the body that conforms to an idea understood by the mind or perceived by the senses."[4]

To ease the frustrating struggle of imagining infinity, Aristotle distinguished between the notions of actual infinity and potential infinity. He said that the collection of natural numbers is potentially infinite because it is inexhaustible—there is no largest number. Hence, to deal with infinity, it is not necessary to imagine infinity—all that is needed is an imagined inexhaustible process. It is difficult to logically grapple with the cognitive image of the result of a never-ending process. And Hume tells us to look at an ink spot:

> Put a spot of ink upon paper, fix your eyes upon that spot,
> and retire to such a distance, that at last you lose sight of it;
> 'tis plain, that the moment before it vanish'd the image or
> impression was perfectly indivisible. 'Tis not for want of rays
> of light striking on our eyes that the minute parts of distant
> bodies convey not any sensible impression; but because they
> are remov'd beyond that distance, at which their impressions
> were reduc'd to a *minimum*, and were incapable of any far-
> ther diminution.[5]

Aristotle believed that scientific knowledge comes from logical
inferences built from indisputable, self-evident truths, but one
might ask where those self-evident truths come from and how
humans ever get to know these truths. Aristotle's answer is that we
induce them. He defined *induction* as acquiring knowledge from our
senses. Aristotle is responsible for making induction the dominant
method of scientific investigation for two thousand years.

Scientific methods changed, however, when investigators in the
seventeenth century suggested that there could be two different
kinds of sciences, *rational* and *empirical*. The term *rational science*
has come to refer to statements coming from ideas, while *empirical
science* refers to, or uses, statements that are based on observations.
Galileo inferred from seeing sunspots that the sun revolved; how's
that for a mixture of reason and observation? It is now reasonable to
ask the Marx Brothers' question, "Who'a ya gonna believe—me or
your own eyes?"

Rational science is about ideas, whereas empirical science is
about experiences connected to the real world. Rational science can
tell nothing more than the relations between ideas, while empirical
science can describe the properties of real things. Rational science
starts from so-called self-evident truths and moves by rigorous logical arguments to a conclusion about some truth; as we've seen, this
process is called *deduction*. Empirical science starts from a hypothesis, which implies a tentative truth, and uses multiple observations to
reach a conclusion about the truth of the original hypothesis. This

process of arriving at the truth is called *induction*.

Mathematics is a rational science, and yet one of the axioms of arithmetic feels strangely inductive. Though the axiom is called *the induction principle*, it actually has nothing to do with the scientific method of induction. To use it, one must make deductions from hypothetical truths. It then draws a general conclusion from the specific deductions. We might call it mathematical induction, though it is not the same kind of induction we see in modern methods of scientific investigations.

How is mathematical induction different from scientific induction? The degree of certainty in scientific induction depends on the number of observations and confirmations, while mathematical induction implies absolute certainty—well, as absolute as anything can be. One might be convinced by a scientific induction argument, but never absolutely certain without observing an infinite number of cases. But a mathematical statement is accepted only as a logical consequence of already accepted mathematical statements. It is not accepted simply as a matter of checking a large sample of reductions to specific cases. We use scientific induction to learn and discover; we start with our immediate experiences and with what we know, and proceed from observation to observation with the hope of approaching a more universal understanding.

The induction principle can be interpreted visually as an infinite line of dominoes standing on edge spaced less apart than their height. This would ensure that if any domino fell toward its neighbor, it would knock down its neighbor.

This alone would not tell us that all the dominoes must fall. But if we knew that, say, the one hundredth fell, then we would know that all the dominoes to the right of the ninety-ninth must fall. And if we knew that the first fell, we would know that all must fall.

Now, this domino metaphor gives some highly intelligent thinkers trouble because it requires thinking in the real world, where dominoes take time to fall; it would take an infinite amount of time for all the dominoes to fall. One of my students wryly defended his position of not accepting the induction principle by calling my attention to a domino-toppling setup done back in 1984 by Klaus

Friedrich in what was then West Germany. Klaus set up 320,236 dominoes in a very complex formation and gave the lead domino a push. The falling stopped after almost thirteen minutes with 281,581 dominoes down, leaving 38,655 dominoes standing.[6] But the mathematical line of dominoes has no physical existence; it is merely an ordered list of statements, each one on the list relaying validation to its neighbor. Here's how it works: Suppose that we have some statement that depends on a whole number n—label it $S(n)$. For example, suppose that $S(n)$ is the statement

$$1 + 2 + 3 + \ldots + n = \frac{n(n+1)}{2}$$

So, $S(n)$ is represented by the nth domino. This means that

$S(1)$ says that $1 = \dfrac{1(1+1)}{2}$

$S(2)$ says that $1 + 2 = \dfrac{2(2+1)}{2}$

$S(3)$ says that $1 + 2 + 3 = \dfrac{3(3+1)}{2}$

Now, suppose we know that the statement $S(n+1)$ follows from statement $S(n)$ for every positive whole number n. In our metaphor, this is represented as the nth domino falling and hitting the $(n+1)$st. Further suppose that we know that $S(1)$ is true. Then the domino metaphor suggests that $S(n)$ is true for every positive whole number n. This is the idea behind one of the axioms of arithmetic that permit us to prove things about infinite sets. It is called the *induction principle*.[7] Though it is now established as an axiom, it originated in the sixteenth century as a technique for proving theorems that had already been suspected as true.

As we shall see, a statement such as "$1 + 2 + 3 + \ldots + n = n(n + 1)/2$" can be proven without using the induction principle, but there are others that unavoidably use it. For example, it would be difficult (though not impossible) to prove that the sum of the first n cubes of integers is equal to the square of the sum of the first n integers. The induction principle proves it very quickly.

A student came to my office one day to tell me that someone had stolen his calculator and that that was his excuse for not having completed the previous day's assignment. He was very upset because he had lost all his stored programs and could hardly function without them. I tried to reassure him that his calculator would turn up. During our conversation, he boldly confessed that he didn't believe a proof that I had explained in class that day. We had just started learning a topic of calculus that uses the formula $1 + 2 + 3 + \ldots + n = n(n + 1)/2$.

"It looks like it's true for all the numbers I can test it on," he said. "How can you know that it's true for all numbers?"

I quickly responded, "Aha! Let me show you three different proofs. Afterward, you will have to tell me which, if any, you believe." I spent the next hour writing the three proofs on a blackboard and explaining each one. I explained the idea behind the induction principle and was surprised that he seemed to accept the axiom. I then gave him a proof based on an infinite construction.[8]

Two schools of convention have opposing views on this last proof. For Platonists, the whole infinite list of statements already exists together with their truth values, regardless of whether humans validate them. So, the Platonist has no problem with the notion that we can show that all statements on the infinite list are validated by virtue of the general link between the nth statement and the $(n+1)$st. But the constructivist cannot accept this because each statement must be validated, in turn, and because all mathematical statements are invented, they must be brought into existence one by one. Platonism and constructivism are not the only possible views; rest assured that these two opposing views split into many hair-splitting refinements, some being very current and trendy, but I will not get into them.

The two other proofs do not appear to use the induction principle. The second, given below, comes to us by way of a story about Carl Friedrich Gauss (1777–1855). Gauss was ten years old when he was admitted to a class in arithmetic taught in an overcrowded, neglected school in Braunschweig, Germany, by a vicious

Herr Professor Büttner, who was known to brutally flog the students for little reason. As a class punishment, Büttner assigned the time-consuming exercise of summing all of the first one hundred whole numbers. (Actually, the problem was far harder than the one listed here, but summing the first one hundred whole numbers will do for the anecdote.) The young Carl Friedrich instantly wrote a single number on his slate and threw it onto the table with the correct answer before Büttner finished stating the problem.

"Lieget se," he said. ("There it lies.")

How did he do it? He simply noticed that if he wrote the numbers twice, once in ascending order and once again in descending order, he could add the numbers in pairs (1+100, 2+99, 3+98, and so on), always getting 101. Because there were 100 pairs and twice as many numbers as he wanted, his final answer should have been $(101)(100)/2 = 5050$. Of course, Gauss's idea generalizes to give another proof of the fact that $1 + 2 + 3 + \ldots + n = n(n + 1)/2$.

I explained the second proof, what I would call a finite algebraic proof.[9] Finally, I presented my third and last proof, a variation on the picture proof referred to in Chapter 4. I thought I could convince my student by using some geometry.

Suppose that we arrange objects (black dots) in a triangular way (figure below) and try to count the total number of objects (black dots) in the triangle.

Arrange the same objects to make a rectangle (black and gray dots) of sides n and $n+1$.

You now have $n(n+1)$ objects. But you can see from the figure that you want only half the objects (only the black dots). So, the number of objects (black dots) must be $n(n + 1)/2$.

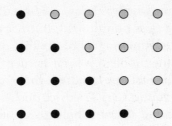

At this point, I stopped to ask my student whether any of this made any sense to him. As I suspected, he would not accept the induction argument, even though I thought he had understood it.

"What about the second proof?" I asked.

"The second proof makes sense only when you specify the last number of your sum. You can't say that it works for any number!" he answered emphatically.

"But the same proof will work no matter what the last number is," I retorted, still thinking that I knew which proof he would accept.

"No, no, I like the last proof," he said with a bit of glee that made me think he was just humoring me.

"And why is that?" I asked, ready for the catch.

"Because it's clear that whatever n is, you just have to count the dots in the picture of the triangle to see that you must have $n(n + 1)/2$," he answered.

Now, here is what is strange about his answer: The figure applies only to $n = 4$. But our minds could have used the figure as an imaginary icon that showed the sum for any n. The process did not depend on the number of black dots in the figure. It would have been the same for any number of black dots. Isn't it extraordinary that we unwittingly projected our proof from the finite case to the infinite? It was natural—as natural as the induction principle. We actually used the induction principle without realizing it. We projected the idea that if it were true for four dots, it would also be true for five. In fact, a central piece of the argument is that we were viewing four dots while thinking that four was arbitrary, and, therefore, we projected the idea that if it were true for n dots, it would also be true for $n+1$.

The real difference is that induction requires knowledge of the formula before beginning the proof. But the geometric arrangement of dots implicitly suggests the formula.

I discovered that my student was confusing mathematical induction with scientific induction, so I explained the difference to him.

Inductive reasoning in science is very different. Unlike mathematical induction, scientific induction leads only to a probable truth. And things can go very wrong with probable truth. Take Carl Hempel's humorous inductive argument that "all swans are black." A logically equivalent statement is "All things that are not black are things that are not swans." But to show that that statement is valid by scientific induction, we have to find only a large number of things that are not black and not swans. There are billions of things that are not black and not swans. We look around and see thousands of objects that are not black and not swans. Each new such object brings us closer to the conviction that "all swans are black."

This is an inductive argument. But it is different from the inductive argument that showed that $1 + 2 + 3 + \ldots + n = n(n + 1)/2$.

For one thing, we know that most swans are white. Surely, you believe that there are white swans out there in the real world, just as you believe that the world is round and not a drum floating on water. But you have never seen a number n that invalidates the previous formula.

Another difference is that objects of the real world are not ordered, but the natural numbers are. When you look for objects that are not black and not swan, there is no ordered sequence to be followed. And because there is no order to your search, there is no inference from one to the next because there is no "next." In fact, all the white swans of the world could be resting on some island that you will never see. You might go through a lifetime of examining objects from every corner of the world, except for the only island habitat that has the exiled white swans.

One more difference is that the set of objects in the real world is finite, while the set of natural numbers is infinite. This is very important. If a white swan exists in the world, it could eventually be discovered through an exhaustive search for all objects that are not black. But if there is a natural number n for which the formula is invalid, it can never be discovered by a sequential search because you cannot exhaust your search through the infinite set. You would be looking for a single needle in an infinite stack of hay.

The real difference is in the inferential part of the induction principle, which has no counterpart for black objects of the real world. We prove that "if $S(n)$, then $S(n+1)$," where $S(n)$ is the statement of the previously given formula. If nonblack objects were ordered, we would have a list of nonblack objects. Then we could let $S(n)$ be the statement "The nth nonblack object on the list is not a swan." But we now see the real problem. We have no way of making the logical connection that relates $S(n)$ with $S(n+1)$. The great jump from "some" to "all" is something that requires an ordered sequence in a deductive system. The missing connection that logically relates $S(n+1)$ to $S(n)$ is a deductive part of the induction principle. In other words, it is here that one must deduce that $S(n+1)$ is true whenever $S(n)$ is true.

In some cases, objects of scientific study can be ordered. We might want to observe something about the nature of temperature and its relation to pressure at a fixed volume. The situation might be that you have left a closed soda bottle in your car on a hot, sunny summer day and are concerned that the bottle will explode.

Fortunately, someone else studied the relationship of temperature, pressure, and volume and found that, for a fixed volume, such as that of the soda bottle in your car, the pressure varies directly with temperature. It is a law of physics called *Boyle's law*. To be convinced that Boyle's law is true, you must conduct an inductive test. In an orderly way, you can test that the ratio of temperature to pressure is always a constant. You could observe this by experimenting with the proper instruments. You would become more convinced that the law is true as the number of observations increased. The degree of persuasion would depend on the number of single confirming observations you made. But there would be many temperature observations that would be missed; after all, there are infinitely many temperatures between any two observed temperatures. You could not observe all temperatures and corresponding pressures, but you could observe enough to persuade yourself that Boyle's law is true.

Our mathematical induction principle is based on the idea that the natural numbers are discrete, following an increasing order from one to the next. It would not apply to Boyle's law because temperatures run the gamut of real numbers, which do not discretely jump from one to the next—for any particular real number, there is no *next* real number.

CHAPTER 11

All Else Is the Work of Man

The Surprising Arguments of Set Theory

Is it ourself, our mind or spirit, that is infinity's proper home? Or might the infinite be neither out there nor in here but only in language, a pretty conceit of poetry?

—*Robert and Ellen Kaplan,* The Art of the Infinite

B y the late nineteenth century, the mysteries of the infinite were ubiquitous. One didn't have to wander far from elementary mathematics to find unexpected, wacky propositions whenever infinity poked its mischievous spirits into the act. How could it be that there are *as many* even numbers as all numbers combined, both even and odd? Surely the collection of all even numbers must be smaller than the whole collection of all numbers! But Galileo, back in the seventeenth century, had shown that that was not so. How could it be that there are just as many points on a line of length one inch as there are on a line of length one foot? Bernard Bolzano had proven it early in the nineteenth century. It's still stranger to find that the *number* of points enclosed by a square (or even a cube) is the same as the *number* of points on just one of its sides.[1]

Some time ago, roughly ninety million years ago, the continents of Africa, Europe, and Asia collided to form a great mountain system. Three-hundred-and-fifty-million-year-old bedrock slowly lifted, folded, and crushed together to form the Alps. In time, melting snows formed two glacial lakes, preparing the sight for one of the most beautiful spots in the world, the small town of Interlaken. That earth-shattering preparation of high country pastures led to another momentous event. Late in the summer of 1872, two young men met while on holiday in Interlaken and became good friends: Georg Cantor, Extraordinary Professor at Halle, Germany; and Richard Dedekind, Professor of Mathematics at the Braunschweig Polytechnikum, a hundred miles from Halle.

Interlaken was a favorite tourist town for Europeans and not a terribly long trip for Germans, even in the nineteenth century. So, the two men would summer in Interlaken and discuss ideas of set theory, perhaps as they explored medieval castles and hiked small trails around the two picturesque lakes. In the middle of August 1874, when singing finches were out in great numbers and the town's groomed parks were abundantly filled with color, the two men would stroll and talk about infinity. One can only imagine their interactions in Interlaken, and how inspired they must have been as they pleasantly promenaded along the banks of the river Aare, strongly aware of the panoramas of mighty Alps with their snow-capped peaks puncturing clouds four thousand meters above them.

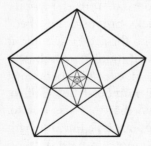

They were not the inventors of infinity. That distinction goes to the early Greek philosophers, who stumbled on its powers when they discovered how to construct the regular pentagon, the five-sided figure that leads to an infinite nest of shrinking pentagons and an infinite expansion of growing pentagons. Nor were they the discoverers of infinity's bizarre implications, for Zeno, as we've seen, was the first to warn us about them. But they were on the verge of discovering some of infinity's most private secrets, along with some of its most concealed bizarre idiosyncrasies.

As the two mathematicians walked for two miles along the Aare until the river lost itself in Lake Thun, they talked math while sensing the proximity of the majestic snow-capped Mount Junfrau, 4,158 meters into the sky. They planned to explore the ruins of one of Interlaken's twelfth-century castles where the mountains shelter the valley from the north winds, permitting an unusually mild climate to form meadows of wild flowers and groves of hardy walnut trees. Absorbed in conversation, they continued up a winding road through walnut meadows to the remains of Unspunnen Castle. Its two towers were intact, and, of course, the two mathematicians debated why one was circular and the other square.

Fir saplings rooted between the stones of each tower were finding their way to sunlight. Cantor and Dedekind sat in the narrow passageway connecting the two towers that influenced a bend in conversation toward the ancient problem of squaring the circle, a problem that would not be fully solved for another eight years.

The problem was simple to state. Given a circle enclosing an area equal to A, could a square be constructed whose enclosed area is also equal to A? (The only tools permitted in the construction are a straight edge and compass because Euclid's postulates translated into those tools: *Two points determine a line that can be extended indefinitely, and a circle of any radius can be centered anywhere.*) This had always been known as the problem of *squaring the circle*. For example, the circle having radius 1 encloses an area $A = \pi$. So, for a square to have the same area, it must have sides equal to $\sqrt{\pi}$.

This problem baffled mathematicians since the fourth century B.C., thirteen hundred years before the invention of algebra, when clever Greeks found geometric schemes to compute enclosed areas by constructing rectangles enclosing the same areas.[2] So, given a figure enclosing some area, can one construct a rectangle enclosing the same area using only a straight edge and a compass? If so, the area of the figure is easily calculated by simply multiplying the length of the rectangle by its width. A natural question follows: For which figures can one construct a rectangle of equal area? Is the circle one of them?

Cantor and Dedekind knew that π was irrational, but that did not give an answer to the problem. After all, even the Pythagoreans knew that $\sqrt{2}$ is irrational[3] and that it is certainly constructible using a straight edge to connect opposite corners of a square of length 1. To solve the problem, one had to show that there is some polynomial equation whose solution is π.[4]

Cantor and Dedekind sat for hours between those circular and square towers. Who knows what direction their conversation took? Perhaps they sat in the passageway, facing each other like a pair of Rodin's *Thinkers* with elbows on knees and hands playing with the curly whiskers of their thick beards. They might have debated whether π is *algebraic* or *transcendental*. (An *algebraic* number is a number that is the solution to some polynomial equation with integer coefficients. A *transcendental* number is a number that is not algebraic.) If it were transcendental, there would be no hope of squaring the circle.[5]

Cantor already knew that there were bucket loads of transcendental numbers—even that they were infinite. He was aware of Galileo's argument that there are *just as many* even numbers as all numbers, and of Bolzano's argument that there are *just as many* points on a line of length one inch as there are on a line of length one foot. And Cantor himself had already proven that there are *just as many* rational numbers as there are integers (see Appendix 4).

Nobody knows for sure what the two mathematicians talked about during their many walks along gravel roads winding through walnut forests and meadows near Interlaken, but their research overlapped remarkably. Both were interested in the real number line— that is, the numbers represented by the points on a mathematical line. Such a line can be understood only through imagination, but an illustration might help the mind comprehend the line.

Draw a line interval whose length is one unit. Measuring from left to right, the first point on the left edge of the interval represents the number 0; a point at distance *d* units (less than one unit) from the left edge represents the number *d*. In the illustration below, the point we are calling 1/4 is one quarter of a unit from 0, the point we are calling $\pi/4$ measures $\pi/4$ units from 0. A point *p* is called a *rational point on the interval* if *p* is a point whose distance from the left end is a rational number of units. *Integer point, irrational point, algebraic point*, and *transcendental point* are defined in a similar way.

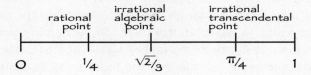

The big question for Cantor and Dedekind was this: If there are *just as many* rational numbers as there are integers, then shouldn't one expect the same for irrational numbers? After all, between any two rational numbers there is an irrational, and between any two irrational numbers there is a rational. (See Appendix 3 for the reason.) And the same question arose with algebraic numbers: How frequently did they appear in an interval? The question refers back to what *just as many* means in reference to infinite collections of numbers. Does it mean that the two collections have the same number of numbers? If it does, there must be some way to *count* the numbers. We know what it means to count the members of a finite collection of objects. If we want to count strawberries in a quart container, we associate with each strawberry a whole number in sequence: The first would be associated with 1, the second with 2, and so on until the last strawberry is counted. If the last strawberry is associated with the number 12, we say that there are 12 strawberries in the container. But what could it mean to count the members of an infinite collection of objects, an infinite basket of strawberries?

Cantor had a brilliant idea. He had already invented the notion of a *set*. A set is simply a collection of things (*members* of the set). By *thing*, he meant anything. Examples? All strawberries in a basket form a set. All integers form a set. All rational numbers form a set. To convince ourselves that there are just as many integers as there are rational numbers, we need only to correspond each and every rational number with one and only one integer. And Cantor had brilliantly constructed such a correspondence.

In ordinary language, it is reasonable to use the word *count* to suggest a way to find the number of objects in a collection. For example, someone might say that he or she counted a dozen strawberries in a basket. One might even say—although it is unlikely that one would—that the set of strawberries in that basket is *countable*. This would be a rather informal use of the word, but the meaning would be clear. It would mean that there is some number associated with the amount of strawberries in the basket. But a mathematician's language has to be far more formal and based on strict, unambiguous definitions. Mathematicians say that a set is countable if there is some correspondence that associates one and only one integer with one and only one member of the set.

By the time of their summer walk together, Cantor and Dedekind knew that the set of all rational numbers is countable. But logic and intuition in the finite world do not always apply in the infinite. The most bizarre things happen when infinity is at play. If there is an irrational between any two rationals, and a rational between any two irrationals, we should expect that there would be just as many irrational numbers as rational numbers. And surely it must have been surprising to find out that the set of rational numbers is no bigger than the set of integers; after all, between each integer there are infinitely many rational numbers! But Cantor found two extraordinarily surprising airtight arguments: one demonstrating that the set of rational numbers is no bigger than the set of integers, and another demonstrating that the set of irrational numbers is far bigger than the set of rationals. In other words, the set of rational numbers is countable, while the set of irrationals is uncountable. (See Appendix 4 for Cantor's demonstrations.) Take a minute to think about how counterintuitive these arguments are. We have a collection of two sets of numbers that go on forever. Between any two members of one set there is a member of the other, and yet one of those sets is vastly larger than the other.

The set of real numbers (those numbers that can be expressed as a—possibly infinite—decimal expansion) can be represented on the number line by the set of points extending infinitely in two directions from a fixed point that we call 0. Of course, this set can also be represented by the decimal representation of a number. Here are some examples: 2, 2.5, −3.33… (the three dots indicate that the 3s repeat forever), 3.141592654... (here, the three dots indicate that the digits continue forever, with no repeating pattern). If the set of real numbers is bigger than the set of rational numbers, it is natural to ask: How much bigger? What does bigger mean when both sets are infinite? To answer this last question, we define the size of a set by its *cardinality*. The number of members in a set S is called the *cardinal number of S* or the *cardinality* of S. If S is a set with three members, its cardinality is 3. Now, of course, this seems to make little sense when S is infinite; in that case, the cardinal number of S cannot be a number in the usual sense of number. But we can distinguish sets of different sizes by considering the question of whether they can be put in one-to-one correspondence with each other. For example, the integers and the rational numbers are the same size. If a set A can be put in a one-to-one correspondence with a subset of a set B, but not with B itself, it would be reasonable to say that B is bigger than A.

If a set is infinite, you cannot make it bigger by simply throwing in a finite bunch of numbers. In fact, it's possible that you might not make it any bigger by throwing in even an infinite bunch. However, there is a way of making it bigger. Here's how: Start with the set S and consider the set of all its subsets (all sets that can be made up from the members of S). This new set is called the *power set* of S, labeled as $\pi(S)$. For example, take the set consisting of the letters a, b, and c. Its power set is the set consisting of eight members: a, b, c, (a, b), (a, c), (b, c), (a, b, c), and one other interesting member called *the empty set*, labeled as $\emptyset$. The empty set $\emptyset$ is the unique set that contains no members. It is easy to show that the empty set is a subset of any set.[6] To show that $\emptyset$ is a subset of S, all one has to do is show that every member of $\emptyset$ is also a member of S. However, there are no members of the empty set and, hence, no members to check. Put another way, we have checked all the members and, consequentially, have shown that the empty set is a subset of every set.

Naturally, the power set $\pi(S)$ of a finite set S is always bigger than S because each member of S can be considered as a subset. But a powerful thing happens when S is infinite. It turns out that the power set $\pi(S)$ is always bigger than S in the following sense: There is no one-to-one correspondence between the power set of S and S

itself. This means that there is a hierarchy of infinities just by letting S be the set of integers and considering $\pi(S)$, $\pi(\pi(S))$, $\pi(\pi(\pi(S)))$, and so on. This hierarchy comes from lining up the cardinalities of these larger and larger power sets.

This means that there is at least a countable number of sets (with ever-increasing cardinalities) such that no two are in one-to-one correspondence. Infinite sets could be distinguished by the sizes of their infinities. If aleph-zero is the smallest infinite cardinal number (the cardinality of the integers) and c is the cardinality of the real numbers (the continuum), then we know that aleph-zero $< c$. Cantor believed that there are no cardinalities between aleph-zero and c, but he could not prove it.

From where they sat, they could not see the Finstevaarhorn dramatically rising 4,274 meters into the sky. But the exceptional panorama of the valley, the high-country pastures with their grazing herds of cattle and the shorter high mountains, was spectacular enough to inspire transcendent thoughts. It was that same panorama that had inspired Lord Byron to set his play *Manfred* at Unspunnen Castle. On the walk back to Interlaken, the two mathematicians discussed the question of whether there is a cardinal number between aleph-one and c. That question haunted Cantor for the rest of his life. One day he thought he had proven the continuum hypothesis. But on the very next day, he thought he had proven it false. His attempts would alternate between believing that the continuum hypothesis was true and believing it false. Day by day, his opinions would change, but in the end, he hypothesized that it was true. This hypothesis became eminently known as the *continuum hypothesis*. When he first posed this question, he could not have known—or even dreamed of—the surprising answer. Half a century after his death, the world would know that it is impossible to know the answer to his question.

What does it mean to say that it is "impossible to know the answer?" Ever? Couldn't some exceptionally gifted prodigy come along a thousand years from now with an ingenious way of finding an answer? After all, there are only two possible answers—yes and no—and it's one or the other. Or is it? In pondering the answer, we are led to new questions for which we have several possibilities. What if it is impossible to know the answer? Could there be a question without an answer?

It's possible that there is no logical connection between the continuum hypothesis and the other axioms of arithmetic. A proof of the continuum hypothesis would require a sequence of logical deductions leading from the axioms of arithmetic to the continuum hypothesis. If a proof is not possible, we should expect the continuum hypothesis to be, at the very least, consistent with the axioms of arithmetic; meaning that its truth does not contradict any of the axioms. At most, the continuum hypothesis might be independent of axioms of arithmetic. This would mean that it has no logical connection to the axioms of arithmetic.

What could we conclude then? Could we possibly conclude that the answer is both yes *and* no? Sure. But we would demand a mathematical proof for either answer—yes or no. And such a proof would require the mathematician to deduce the answer from a sequence of logical deductions indirectly linking back to a collection of axioms. In particular, because the question involves the real numbers, the collection we mean here is one that contains the axioms of arithmetic. That collection—as well as any collection of axioms—must obey at least one condition: The axioms in the collection must be logically independent of each other, meaning that one cannot be logically deduced from the others. This condition guarantees that future deductions from the collection will never lead to contradictions. But could the collection be incomplete, meaning that there might be some new statement expressible in the language of arithmetic that is independent of the other axioms already in the collection? Such a statement could never be proven by deductions from the axioms. So, couldn't the continuum hypothesis be a candidate for one of those axioms? If it is, there would be no logical connection between the axioms of arithmetic and the question of whether the set of real numbers is just as large as aleph-one. Hence, a yes or no answer would make no sense. One could add a new axiom to the collection. One could add the continuum hypothesis: There are no cardinalities between aleph-zero and c. Then, of course, the continuum hypothesis would be true simply by virtue of being an axiom itself. But one could just as well add the negation of the continuum hypothesis to the collection in place of the continuum hypothesis and never encounter a contradiction.

In the spring of 1963, the answer was put to rest. Paul Cohen, a young Stanford mathematician, used a 1937 discovery of Kurt Gödel's to prove that the continuum hypothesis is independent of the axioms of arithmetic. The answer to Cantor's question is yes *and* no. "In 'Cantor's Paradise'," writes Simone Weil in her *Notebooks*, "the mind has got to be very much clearer, more exact and intuitive than anywhere else (as in the mysteries of theology.)"[7]

Imagine the horizontal line segment between 0 and 1. Cantor's shocking revelation that there are far more irrational points on the segment than rational points means that, if only rational numbers are visible, the number line has gaping holes everywhere, so it is far from being a solid line. We cannot ignore those irrational points; after all, they correspond to true numbers. Look at it another way: Ignore all the rational points, and you have a line with far fewer holes. In fact, there are so few holes that it would be impossible to hit one by randomly striking it with a vertical line (that is a mathematical line). The probability of hitting a hole is absolute zero; you will always hit one of the irrational points. Once again, infinity's logic throws intuition a curve ball. The set of irrational numbers is huge compared to the set of rational numbers. Perhaps it would be more understandable if we were to say that the chances of hitting a rational are infinitesimal, given the huge difference in sizes of the two sets. However, the probability that an event will happen is some number between 0 and 1, so it would be meaningless to say that a probability is infinitesimal and not assign it a number. The only possible candidate for such a probability is the number 0.

There is another way to look at this. Imagine a circular dartboard. Take out all the points that are a rational distance from the center, so you are left with an infinite number of circular gaps.

(The radii of those gaps are rational numbers.) Now throw darts at the board trying to hit one of the circular gaps. You never will. Of course, we are assuming that your darts are mathematical darts, the kind that only the mind can manufacture—the kind with points of dimension zero. If you throw any number of such mathematical darts, the probability of hitting a circular gap is zero! You might feel that you can hit one if you are given a large number of tries. Perhaps you feel that you need infinitely many tries. But when the probability of hitting a rational circle on one try is zero, that probability stays zero for any number of tries, even infinitely many.

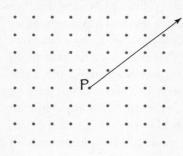

Here is yet another way of looking at the phenomenon. Imagine a grid of points spaced integer distances apart both horizontally and vertically. Place an infinite straight line on the grid in such a way that one end is at one of the points of the grid—say, at *P*. Then imagine that the line is free to rotate about *P*; so, in effect, we have a spinning dial. Now spin the dial and assume that there is some strange perfectly random friction that slows the dial to indiscriminately stop and point in some direction. If someone offers you a million-to-one odds on hitting one of the points of the grid (other than the point fixed to the spinner) would you be willing to bet a dollar?

Save your dollar. To win, your spinner would have to come to either rest on a vertical line through *P* (an infinitely rare event) or else rest on a point *Q* that is a whole number of spaces vertically up or down from *P* and a whole number of spaces to the left or right of *P*. That would mean the slope of the line (defined to be the number of spaces *Q* is up or down from *P*, divided by the number of spaces *Q* is to the right or left of *P*) would have to be a rational number. The set of all rational slopes is so insignificant compared to the set of all irrational slopes that you would have zero chance of getting a rational slope.

You were asked to do the impossible task of imagining the line segment between 0 and 1. All you can really do is symbolize it. Try as hard as you can to visualize the effect of having an irrational between every two rationals and a rational between every two irrationals. Hold that image and include the property of having more irrationals than rationals. Recall that a chiliagon is a polygon with one thousand sides. The symbol that you conjure up is in no way different from the one representing a myriagon (a polygon of ten thousand sides) or a hectagon (a polygon of only one hundred sides). You can mentally conjure up a symbol for the number line, just as you construct a symbol for a chiliagon as soon as you try to visualize it.

The fifteenth-century German cardinal, philosopher, and mathematician Nicholas de Cusa saw the circle as the end result of regular polygons forever increasing their numbers of sides. Being a cardinal, he naturally used that image to imply that one can approach truth without reaching it, just as one can approach squaring the circle without ever completely succeeding. Surely, de Cusa was not the first to imagine the circle in that way. Aristotle mentions a mathematician by the name Bryson of Heraclea who lived in the middle of the fifth century B.C. as one of the early circle squarers. Little is known about Bryson, but it seems that he had the idea of inscribing and circumscribing polygons two hundred years before Archimedes. Because the difference between the inscribed and circumscribed polygons could be made as small as one wishes, simply by using polygons with a large enough number of sides, any polygon between the inscribed and circumscribed polygons will approximate the circle to any prescribed degree of accuracy.

Antiphon the Sophist was quite possibly a contemporary of Bryson. His idea was this: If it is possible to inscribe a regular polygon with a large number of sides in a circle, square the polygon instead. That can be done. Increase the number of sides by doubling, stopping only when the degree of accuracy is satisfied. The constructed squares are approaching a square with the same area as the circle. This idea might have led Eudoxus of Cnidus, a fourth-century mathematician, to invent a method that was appropriately called *the method of exhaustion*, a method for approximating areas enclosed by special classes of closed curves. Eudoxus is well known for this contribution.

But his work spills into the broader concern of comparing lengths that might be irrational and permitting acceptance of the real numbers. Thomas Huxley tells us, "It is difficult to exaggerate the significance of [Eudoxus'] theory, for it amounts to a rigorous definition of real number. Number theory was allowed to advance again, after the paralysis imposed on it by the Pythagorean discovery of irrationals, to the inestimable benefit of all subsequent mathematics."[8]

Defining and understanding the set of real numbers was a primary concern of our friends Cantor and Dedekind. Dedekind himself formally defined what a real number really is when the two men met for the first time while on holiday in Interlaken. Though real numbers were used for centuries, it was the first time its illusive spirit was captured in a language that defined it. But not everyone was accepting of Cantor's strange discoveries about the properties of real numbers. Leopold Kronecker, another contemporary with Cantor, took an adversarial view, requiring that mathematical proofs should be deductions with, at most, a finite number of steps using only finite sets. He did not believe that transcendental numbers exist and once made this infamous remark: "God created the integers; all else is the work of man." It is likely that Cantor and Dedekind would have put this distinguished remark the other way around.

Jorge Luis Borges tells a story about the absence of time. He recalled visiting a place just beyond the neighborhood of his childhood:

> "I stood there looking at that simplicity," he writes. "I thought, no doubt aloud, 'this is the same as it was thirty years ago.' … The easy thought, 'I am in the eighteen hundreds' ceased to be a few careless words and deepened into reality. I felt dead—that I was an abstract perceiver of the world; I felt an undefined fear imbued with knowledge, the supreme clarity of metaphysics. No, I did not believe I had traveled across the presumptive waters of Time; rather I suspected I was the possessor of the reticent or absent meaning of the inconceivable word *eternity*. Only later was I able to define that meaning."[9]

I visited Interlaken on an Easter vacation when I was still a student in Paris. The imposing Jungfrau, Schreckhorn, and Finstevaarhorn majestically filled the vistas along my walk through the countryside on a winding gravel road. I passed several stone foundations of houses with tree roots lifting and moving whole walls that were centuries old. Without a destination, I followed the finches as they stopped to rest and scavenge in the fields. Each time I came close, they would move farther along to a new spot until we converged at a ruin on a wooded hillside. I recognized that stone ruin and thought, "I've been here before." And there, between the circular and square towers of the ruin, were two bearded men talking, sitting on stones with elbows on knees and hands on chins.

Perhaps Pythagoras was right to believe that number was the source of all things. It has survived many crises, like the one it faced when the Pythagoreans discovered that the diagonal of a square of side one is irrational and then again when it met the strange paradoxes of the infinite.[10] We demand more and more from number. It must furnish the physicist and engineer with rational approximations to any degree of accuracy so he or she can make predictions in the real world from interconnected events such as butterflies flying over the Pacific and hurricanes in the Atlantic. Even the best approximations that are possible without the logic of infinity, those differing from their true values by astoundingly small amounts (such as 10^{-n}, where n is the number of atoms in the universe), are not good enough to establish some significant scientific theories. Take Einstein's famous equation $E = mc^2$, which made a great deal of new technology possible, including nuclear power and the atomic bomb. That equation is based on two simple ideas for which an understanding of infinity is essential: First, the mass of a moving particle changes from its rest mass by a factor that can be represented as an infinite sum of numbers that depend on the velocity of the particle itself, and, second, that that infinite sum can be approximated by the energy of the particle. It is crucial to know that the infinite sum converges to a finite quantity that represents the ratio of relative masses. Without an understanding of that convergence we could not understand $E = mc^2$.

Understanding the logic of infinity enabled us to make many huge leaps in the progress of science and technology. But when it comes to the pursuit of truth, the logic of infinity isn't the whole story of the success of science. For a more complete understanding we must consider what makes science tick everyday. What kind of proof do scientists need to move forward? Indeed, what kind of truth do we use to get things done—from voyaging to Saturn to determining what foods are healthy? Disciplined plausible reasoning makes science possible. It is the logic of everyday reality.

PART III

Reality

CHAPTER 12

Does Math Really Reflect the Real World?

Anything new that we learn about the world involves plausible reason-
ing, which is the only kind of reasoning for which we care in everyday
affairs. … Certainly, let us learn proving, but also let us learn guessing.

—*George Polya*, Plausible Reasoning

S ounds of the Big Apple were different forty years ago, when
noisy busses spewed smelly dark smoke into the avenues and
car horns pointlessly blasted at busy intersections. It was a time
when manicured islands separated traffic on upper Broadway. Sidewalks of crossing streets were still lined with tall chestnut trees
protected by ornately sculptured iron gates centered in square gar-
dens of smiling multicolored pansies. On early autumn mornings,
shortly before the clanging clamor of galvanized metal cans of incin-
erator ashes, amid the whines of garbage truck compactors, one
could hear sweet hurried trills of field sparrows or mournful coos of
pigeons. That was before the traffic woke, when avenues were still
wet from sprays of street-washing trucks that forcefully showered
dust and oily grime into rivulets that ran along the curbs into nearby
sewers. It was the hour to notice the island with its yellow marigolds

drooping under the weight of droplets from mists left by the quick-moving street-washers. I would walk aimlessly along Broadway spotting the faint scent of chlorine disinfectant before bouquets of rival smells wafted from delicatessens and flower shops.

On one of those warm autumn mornings, I walked north along Broadway, passing newspaper hawkers stacking papers at corners of busy intersections, getting ready for the morning rush. I had taken a semester off from college to earn money for my studies in Europe by rendering new Long Island houses for sale in the classified section of the *New York Daily News*. My mornings were free.

At the entrance to Columbia University, I decided to follow a random group of summer school students to their classes. I followed close behind a group of three chatty individuals talking about the growing war in Vietnam. They entered a large classroom which had a course schedule posted on its door. First on the schedule was a course entitled Chance. Chance, that was all—no other description. Rows of fixed wooden benches behind long narrow tables served as desks in the windowless room. With the queer feeling that my plan was ordained, I took a seat near an aisle and quickly noticed that the room was sparsely filled, mostly with attractive females. I slid more toward the center of the bench to make room for others who were still arriving. My presence should have been conspicuous in such a small class, but when the room began to fill, predominantly with males, an old professor who looked remarkably like Lenin entered and spoke with a soft French accent. I sat through half the lecture understanding little but became interested in what the old man was saying. I guessed that he must have been quite old—but I was amazed at his highly animated enthusiasm. He was short. His face was red from passionate blazes of excitement over what he was saying. He would foxtrot near the lectern with arms flailing, his eyes widening while his head slowly panned across front rows, stopping only for brief stares at select students. The lecture was about replication of DNA and of how—because of random errors in its replication—new cells grow to be slightly altered from their parents, ready to grow and reproduce or not survive, according to Darwinian natural selection. I was young and astounded by the idea that life is supported by whims of chance and luck.

After class, I slowly walked to the nearest subway, thinking of what the class was about. In those days, the subway line that connected Van Cortlandt Park in the Bronx with Brooklyn was called the IRT line. Standing on the downtown platform, I spotted Lenin, the professor. Surely he was not Lenin, but my next fantasy was that the man was Jacques Hadamard, a renowned French mathematician who had proved the Prime Number Theorem, the great contribution to mathematics that tells us that the number of primes less than n is close to $n/\log_e n$ when n is large.[1] It would not have been so extraordinary to see Professor Hadamard taking a southbound subway, except that he would have been over ninety-five years old at the time.

Small groups of passengers migrated toward the doors of the cars as the train stopped at the station. A youngish, big and tall, bearded man pushed the professor in an attempt to secure one of the last few vacant seats of the car. The professor said nothing and accepted the fact that he would have to stand while holding on to a vertical pole while the train careened from side to side as it made its way to the next station. I thought of announcing to everyone in the car that the old gentleman was Jacques Hadamard, as if everyone would get up at once to give him a seat, but then thought better of it.

The professor, whoever he was, got off at my stop. I stayed close behind him as he ascended the stairs to exit the station and followed him along Eighth Street. When he reached Fifth Avenue he turned, looked directly at me, and, with that same flaming excitement he had in the classroom, said, "Who are you? What were you doing in my class, and why are you following me now?"

I said, "Um, I'm not following you; I live there," and pointed to Washington Square.

"Well, who are you then? You are not one of my regular students," he said, uneasily.

"No," I quickly admitted. "I—I just sat in on your class for fun."

The professor tilted his head and slid his eyeglasses higher on his nose with his index finger. "Now I have three unregistered students in my class," he said. "What am I to do?"

"No—no," I said. "Just today—I—I won't come again." But the

professor gave a scolding look to suggest that if I didn't come, he would be angry.

"Do you know the negro boy who was sitting next to you?" he asked. He used the pejorative noun as everyone did in those days, but went on to praise the lad. I felt awkward standing beside him while he talked and told me that—like me—the "negro" boy had "strayed" down from Harlem and curiously wandered into the classroom.

"But, I tell you, he is the brightest student in that class!" he said.

The professor began to unnerve me. He had to be eighty-five. "Was he barmy? Why was he telling me this?" I thought.

"Tell me," I said. "What is your class about?" It could have been about psychology of gambling, probability, or analysis of scientific reasoning.

Once again, he pushed up his eyeglasses, which had fallen lower on his nose. "Come," he said, waving a finger in the direction of the park. He started walking and I followed. "Tell me, how much mathematics do you know?" he asked.

He was not my professor. He couldn't give me a bad grade. Yet I had trouble telling him what he wanted to hear. All I could say was that I had had a bit of calculus. That was not true. I had spent two years studying almost nothing but mathematics.

Leaves in the park were beginning to change color and fall on that windless bright day of perfectly comfortable temperature.

He said, "Let me tell you what my class is about. At the start of every physical event, chance will delicately direct the event's final destiny, and even the most sensitive perturbation of the start may radically alter its final fate. Jacques Hadamard had this idea when…"

"Jacques Hadamard?" I almost shouted.

"Jacques Hadamard was a great French mathematician who died recently. He proved that, when hitting a billiard ball, extremely sensitive deviations of the angle of the cue stick can cause drastically different results."

"You mean the cue ball might hit or miss the ball it's intended to collide with?"

"No," the professor said, "Getting the cue ball to collide with other balls—how do you call them, object balls?"

"Yes, I think so."

"Well, getting the cue ball to collide with an object ball is relatively easy. I'm assuming that the players are good enough to be able to hit the cue ball head on. It's what happens next that could have wildly different consequences. For example, in one case, the cue ball might hit the 2 ball, which, in turn, might hit the 4 ball into the desired pocket; yet, with the slightest perturbation, the 2 ball might completely miss the 4 ball. To precisely predetermine the initial angle of the cue stick, one must battle all the chances of the minutest conditions that might upset the hit. This is what sports are all about."

"What you're saying is that every time a rack of pool balls is broken, there is a new game, a game that has never been played before," I said.

"I'm saying that the fall of a great kingdom can be attributed to the loss of a single horseshoe nail. Look over there." He pointed to some chess players in the park. "Every time someone makes a move, it opens the game to vastly different subsequent moves."

"That's what your class is about?"

He laughed, "Not exactly. Come to my lectures to find out." He got up and walked back to Fifth Avenue.

Two mornings later, I found myself walking uptown at a fast pace on my way to Columbia University. I bought a blank notebook at the Columbia bookstore and went to Chance. This time I sat next to the young wanderer from Harlem. He had a kind, round face and dressed in a faded yellow T-shirt showing wear from countless washes. His name was Uriah Brown.

We were both there early, with enough time to talk a bit and notice the room while glancing toward the door whenever a woman entered. The number of attractive females interested in this bizarre class surprised me. Six blackboards covered the front wall, three pairs with one sliding over the other. Obscure mathematical fragments covered the dusty boards with a dense collage of symbols and diagrams. The milk glass chandeliers flickered all at once just as the professor entered the room.

Although I had taken a course in probability and statistics, I never had very much understanding or confidence in the material. After only one class session with the professor, I realized that my logic had to shift to a completely different kind of reasoning. He talked about dice. It was a simple lecture about two different kinds of dice. One was a mathematical cube, in which one face was as likely to show as any other—an idealized die. The other was real, made so well that a shake would cause it to roll onto one face as likely as any other.

The professor designed his lesson around an experiment. A large glass bowl of dice was passed around the classroom. The milky glass bowl was remarkably similar to the chandeliers hanging from the ceiling. Each student picked one die from the bowl and tossed it sixty times, recording the number of times it fell on each side.

"We know that the mathematical die will fall with three dots facing up every six rolls," said the professor. "The question is, will that be the case for the real die? The answer depends on the precision of the real die, on how well balanced it is, and on how carefully the manufacturer considered the differences between sides with differing numbers of spots.[2] If we assume that the die is *fair*—that is, symmetrical in all aspects—we should expect that a three would turn up, on average, nearly one sixth of the time. Experiment with a real die. Roll it sixty times. Surely, you would be astounded if it fell on each of its six faces ten times. And you would be more astounded if after six hundred rolls the same would happen. No matter how fairly the die was made, you would not expect it to be in exact agreement with the mathematical die. If you rolled a real die sixty times and three turned up half the time, you would be suspicious that the die was biased to the three."

And there was his point: The model of the idealized die could be used to test the fairness of the real one. In other words, we should assume that a real fair die should behave very much like an idealized mathematical die. We accept a new form of logic when we presume that the practice of the real world behaves as the idealized theory of the mathematical world.

"The ideal die," he went on, "is entirely predictable. It falls every time without bias because it is defined to be unbiased. On the other hand, the real die is subject not only to imperfections of manufacture, but also to sensitive conditions of the throw. Did it roll along one axis as it left the palm of the hand? How did it hit the surface—rolling along the 2, 3, 4, and 5 faces? If the answer is yes, then a 2, a 3, or a 4 is more likely to turn up than a 1 or a 2. What was the thrower thinking of as the die left the palm of his or her hand? Can the thrower's thoughts influence the outcome?"

At this point, a flurry of answers blended into an unintelligible clamor. Several students were trying to challenge any positive answer to that last question. When the noise died, one female student in the front row said, "Do you mean that repeating to yourself 'give me a 2' can make the die favor to the 2?"

"Yes, that's what I mean," said the professor.

"But how can it?" continued the student. "Shouldn't we have to assume that the thrower did not see the original position of the die in his hand? And if the thrower doesn't know the original position, then how in the world could any thoughts influence the outcome?"

"Good thinking," said the professor. "But this hypothesis that a thrower's thoughts can influence the outcome could be tested, could it not? In fact, we could test it ourselves. But before we do, we have quite a bit more to learn about testing hypotheses. First, each of you should try to test your die against the mathematical die. For example, how do you know that your die is fair?"

The same female raised her hand. She asked, "If I throw my die sixty times and each of the six numbers come up nearly ten times, then shouldn't I assume that my die is fair?"

"Very good," said the professor, "But what do you mean by *nearly*? If nine twos came up and eleven threes came up, would that be a good *nearly*? What if eight twos came up? You see, we have a small problem. I said before that you would be very surprised if after sixty rolls each face would have turned up exactly ten times. So you would expect some deviation from what you would expect of throwing an ideal die sixty times, but how much deviation? Toss your die. But before you do, let me tell you that one of you has a die that is biased toward the two. Which one of you has the one biased toward the two?"

Sounds of clattering die rolling followed. The result was a revelation. For many die, the numbers were strangely close to what the idealized model predicted. Others wildly differed from their

predicted outcomes. For those, a bias was clear. Still others had numbers that fell far from the predicted outcome, yet close enough to
make it difficult to decide their fairness.

For four years, I had been trained to use mathematics to attack
questions for their definitive answers. *Nearly* was not a word in my
mathematics vocabulary. I expected *nearly* to mean something
extremely imprecise. To my great surprise, it didn't. By the end of the
class, we had learned that *nearly* meant that after sixty tosses, the
number of twos would have to be greater than fourteen to conclude
that the die was biased toward two. That seemed like a wildly loose
nearly. But I quickly learned and understood the persuasive strength
of that sort of connection of probability to reasoning.

Later, in that same class, the professor had a large glass bowl
filled with go chips. It seemed that the milk glass chandeliers flickered once again as the professor entered the room. I realize I may
have been imagining this.

"There are a thousand go chips in this bowl. How many are
black?" he asked the class.

The bowl was passed around the room. The students were told to
mix the chips, pick twenty chips without looking, call out the number of black chips picked, and then return the chips to the bowl. The
professor drew a horizontal line on the blackboard with equally
spaced marks labeled 1 through 20. As the bowl was passed from student to student, the professor would draw a square box over the
number equaling the number of black chips called out.

"We could sort all the chips and count the black ones," he said.
"But I want you to think of this bowl as a metaphor of a world containing inaccessible measurements—inaccessible, perhaps, because
of expense, complexity, or simply time consumption. Can we guess
at the number of black chips by repeatedly taking samples of just
twenty chips?"

As the bowl was passed, the squares on the blackboard began to
mount in towers higher above the number 6 than above nearby
numbers. Some students called out 4, others 5, and still others
3, 7, and 8. There were even some rare callings of 2 and 9, but predominantly, 6 was the number called.

"What should we expect when we find that we mostly pick six black chips in a handful of twenty?" the professor asked.

"Six out of twenty chips are black," Uriah called out.

"Okay. So, you would automatically suspect that there are three hundred black chips in the bowl."

"Only suspect," Uriah shouted. "Not know for sure."

"Right again—only suspect," said the professor.

Uriah slowly found himself contesting the professor's argument that a handful of chips should reflect proportions of the whole population of go chips. The professor seemed pleased by Uriah's challenge and began to question other students about why they should expect their handfuls of twenty to have nearly six black and fourteen white chips if the bowl contained three hundred black chips and seven hundred white.

"That is what the mathematics says," one student argued. "With a real hand scooping up real go chips, you should expect only nearly six to be black and nearly fourteen to...."

"No," interrupted the professor. "Uriah is right. The law of large numbers says that you should expect the handful of chips to reflect what's in the bowl, provided that the number of chips in the hand is large enough. But what handful is large enough?"

He meant that a handful of twenty chips may show six black chips, but a handful of forty is more likely to show twelve black chips. Increase the handful, and you should expect a more precise reflection of the number of black chips in the bowl. The insightful argument is this: As the handful number increases, it approaches the total number of chips in the bowl. So, if the handful were so big as to scoop up all the chips in the bowl, it would surely contain exactly the number of black chips in the bowl. No black chips in a twenty-chip handful would be a little surprising, but no black chips in a handful of ninety would be astonishing. A handful of ninety should have something in the neighborhood of twenty-seven black chips.

The professor asked students how they would react if a handful of ninety chips turned up no black chips. They understood that although it is possible, it rarely happens and it would be very surprising. By then, Uriah's challenge had steamed up.

"You are talking about chips," he said. "They are real things, not things that can be interpreted through exact arithmetic. I mean, if I flip a real penny, it's never going to be mathematically perfect, so how can you simply assume the math will tell us what it will do ahead of time?"

"Isn't it amazing?" shouted the professor, once again whirling his hands in the air. "You are right. The real world seems to like to stick wonderfully close to mathematics. Taking samples of go chips not only tells us something about the distribution of one color over another in the bowl. But what if the go chips are representatives of something more interesting? Suppose they represent voting residents of New York City, the black ones representing those who will vote for John Lindsay in the upcoming mayoral election. It seems that we can predict the election. It's really a three-candidate election among Lindsay, Beame, and Buckley, so that makes it a bit tricky because no one candidate will get more than 50 percent of the vote. The polls now show that Lindsay has 46 percent of the vote. Beame and Buckley have 41 percent and 13 percent, respectively. How do the pollsters know this? They seem to have the election all wrapped up with only a 3 percent margin of error."

"How, then, can you scoop up a handful of voters?" Uriah asked.

"Aha," the professor responded in his energized manner. "You ask them, but you have to be careful about whom you choose to ask. It's tricky; the accuracy of your prediction will depend entirely on how well you have randomized the scooping. In other words, if your handful of voters is as random as your handful of chips, the election can be predicted by simulation."

Of course, the professor meant that the election could be predicted within a certain degree of error. The number of votes for Lindsay would be just an estimate, but how good should that estimate be? Different handfuls of chips, even though they are scooping up random chips, will show different numbers of black chips. That is expected. But the precision of an estimate depends on the variation of numbers of black chips in repeated handfuls. There is some strange paradox wanting to come out of the box. If repeated handfuls give widely differing results, we would assume a lack of precision. On the other hand, if each of a hundred handfuls of twenty contained six black chips, it would seem that the sampling is giving a precise reading of the proportion of black chips to white—something we would be very suspicious of.

After class, the professor invited three students to have coffee with him. He pointed to me. A young female in the front row and Uriah were the other two. I revised my estimate of the professor's age downward.

He led the way to a luncheonette on Broadway that had a twelve-page menu. We sat for an hour while the professor prompted us to talk about ourselves. I didn't say much. Neither did the girl, Candice. We listened to Uriah. He was a high school dropout who had spent the previous year working at the soda fountain of a drug store in Harlem until it was robbed at gunpoint. He had a beat-up copy of a Modern Library paperback called *Makers of Mathematics* under the counter, which he would peek at during slow times. He might have been nineteen or twenty at the time. One day he ventured beyond the limits of Harlem. He confessed that he had never been south of 120th Street and that there was something in the smells of the early morning air in September beckoning him to wander southwest from his family's apartment near 127th Street and St. Nicholas Avenue. He wandered to a world unlike the one he knew. When he reached the east side of Columbia's campus, he had no idea what was compelling him to enter, but he remembered his copy of *Makers of Mathematics* and searched several buildings for signs of someone teaching mathematics. Like me, he saw Chance written beside a door to a large classroom and entered.

Uriah's addiction to poker and gambling left him helpless in the streets of Harlem, where crapshoots and numbers were common pastimes. He spent a good deal of his weekly paycheck on numbers. But his math thirst made him ask questions about gambling. When he saw the sign on the classroom door saying *Chance* at Columbia, he was bound to enter. What could math teach him about the reality of gambling?

CHAPTER 13

Who's Got a Royal Flush?

Making Predictions with Probability

The probable is that which for the most part happens.

—*Aristotle*, Rhetoric

Uriah continued attending Chance. I stopped after the second class but came to know Uriah by walking a short distance with him after leaving the luncheonette on Broadway. Gardened traffic islands degenerated to neglected weedy loam on our way north; the lonely, isolated chestnut trees looked forsaken. When conversation turned to weightlifting, I invited Uriah to spend an afternoon as my guest at the McBurney YMCA on Fourteenth Street. There, he talked about sports and the chance class, while spotting my embarrassingly light weight bench presses.

Uriah came with me to Le Figaro, my favorite haunt in Greenwich Village. In the mid-1960s, it was easier to get a stash of marijuana than a good cappuccino in New York City. Le Figaro was a mile walk from my gym, but one of the few places where you could get good foam in your cup, take half an hour to sip down to the caramelized raw sugar at the bottom, and still sit satisfied for several more hours, occasionally examining your empty cup, wondering when you drank it dry. Milk glass chandeliers, similar to those in the professor's classroom, lit the high tin-paneled ceiling and cast mellow light through rising cigarette smoke onto walls covered with collages of yellowing editions of the French newspaper, *Le Figaro*, mirrors, framed photos of local poets, and antique menus showing what a nickel could buy in the 1920s.

The professor happened to be sitting near a window smoking a cigarette. Our coincidental encounters were becoming suspicious. In a city that boasts 7.8 million inhabitants spread over 321 square miles, one would think that unplanned meetings would be rare. But there was the professor, sitting at a round table made of small ceramic square tiles, reading a paper over a very large cup of tea. He insisted that we join him and at once cleared his table of crumpled sections of newspaper.

"Hey, professor," Uriah said. He was eager to ask one of his poker questions.

Uriah claimed that he had witnessed a relatively high-stakes poker game in which a royal flush beat out a straight flush. He felt that the winner was cheating and wanted to know the professor's opinion; after all, either flush is rare, so the two together must be exceedingly rare. That was when the professor told us that one deal is as likely as any other.

"The likelihood of drawing an arbitrary poker hand is 1 in 2,598,960," the professor said as his glasses once again slipped down his nose. "One deal is as likely as any other; so why should you be astonished if you were dealt a royal flush?"

"Wait, wait, wait!" Uriah called out. "Man, did you say a royal flush is just as likely to happen as a straight flush? Hey, man, that ain't...."

"Ah, you're right," the professor cut in. "I mean that any one hand is as likely as any other. Four different royal flushes are possible, one for each suit. So the likelihood of getting any one of those royal flushes is 4 in 2,598,960, but the likelihood of getting a particular one—say, a royal flush of diamonds—is still 1 in 2,598,960.

Clearly, it's harder to get a royal flush of diamonds than it is to get an unspecified royal flush."

I took my first look at the caramelized sugar at the bottom of my cup. "You are saying that to get a specific dull hand is also rare, but that there are heaps of possible dull hands, and that we are not surprised when one is dealt."

"But," Uriah said, "if you want to be the one dealt a royal flush, then it must be even rarer. Let's say ten people are in a poker game and you ask what is the probability that a royal flush will be dealt to one of the ten players. Well, that's gotta be much more likely than it will be dealt to a specific player."

"Probabilities," I said. "are relative to the point of view and are not objectively connected to the number of possible hands that can be dealt. It means that the same royal flush can have a probability of 1 in 2,598,960 when viewed as simply being dealt to any one of the ten people playing, but a much smaller probability when we stipulate who should get it."

"That's...," said the professor, who was cut off by flickering lights. With his elbows on the table and fingers of each hand touching, he bounced his thumbs off each other, saying, "Look, what are the odds of finding me in a restaurant or coffee house in this city, if you were to randomly search for me some afternoon? There are, I believe, about eighteen thousand restaurants and luncheonettes and coffee houses in this city, and I could just as well not have been in any one of them, but you just happened to come into this one. Were you surprised to see me sitting here?"

"Eighteen thousand? That's a big number. Really?" Uriah asked. His thoughts were momentarily diverted by surprise at the size of the city.

"I'm just guessing," said the professor. "But were you surprised to see me sitting here?"

"Yes, very," Uriah said.

"Yes, but the likelihood of my being someplace is certain. The likeliness of you being someplace is also certain. The only surprise is the coincidence of you being in the same place as me at the same time. But suppose you didn't see me sitting at this table, or suppose that you went to the coffee house just across the street. You would never have known how close you had come to bumping into me and, therefore, never have been surprised. These near misses happen more than we think."

"Ah, but you live in the neighborhood," I volunteered. "And I often come to this neighborhood." I said this forgetting that I lied to him when he first confronted me on Eighth Street. "So doesn't that increase the likelihood of running into each other?"

"Yes...." It was all he said before the lights went out.

We continued to talk by candlelight as if nothing had happened. Again, Uriah did almost all the talking. He seemed to have an instinct for understanding probability when it came to cards.

"There are only four possible royal flushes," Uriah said. "But there are 36 straight flushes and 624 hands that are four-of-a-kind." He looked directly at the professor and asked, "Right?"

"That's right."

"Well, it might be that the probability of getting a specific royal flush is the same as the probability of getting a specific straight flush, of getting four of a kind, or of getting any other dull hand, but the probability of getting any of the four royal flushes has gotta be smaller than getting any of the thirty-six straight flushes. And it's less likely to get any of the 36 straight flushes than to get any of the 624 four-of-a-kinds."

"Right," said the professor.

His pedagogical method was to start conversations with outlandishly true comments—such as "one deal is as likely as any other," jolting the mind into deep reflection and launching conversation around fiery thoughts.

"Couldn't we treat the cards as we had treated the go chips?" said Uriah.

"Meaning?"

"Mark a black dot on each of the cards you wish were part of your hand. I mean, if you want to draw a royal flush, put a black mark on the five cards: ace, king, queen, jack, ten—all diamonds."

"Ah, then drawing a royal flush is like picking five black go chips from a bucket of fifty-two chips, forty-seven white and five black," said the professor.

"Yea, that's the idea. Isn't the probability of drawing the five black chips the same as drawing a royal flush?"

The professor paused to think. "Yes," he said.

"Well, then, the probability of drawing the first black marked card is five out of fifty-two because you could be drawing any one of the five marked cards."

"Yes."

"But on the next draw, the probability of drawing another black marked card is four out of fifty-two."

"No, no, four out of fifty-one. You removed one card from the deck."

"Right, four out of fifty-one. And on the next draw, it would be three out of fifty. In the end, the probability of drawing exactly five black marked cards is

$$\left(\frac{5}{52}\right)\left(\frac{4}{51}\right)\left(\frac{3}{50}\right)\left(\frac{2}{49}\right)\left(\frac{1}{48}\right)$$

"That's right! And that will calculate out to be 1 in 2,598,960. What you showed is that the probability of drawing a particular royal flush—say, a royal flush of diamonds—is the same as drawing any other hand."

"The chances of putting black dots on exactly the five cards that make a royal flush is also 1 in 2,598,960," I added.

"No, the chance of drawing a royal flush of diamonds is 1 in 2,598,960. The chance of drawing any of the four possible royal flushes is four times better, 1 in…." The professor calculated for a moment, "1 in 649,740."

"It means that you have a better chance of getting a royal flush than of getting the worst possible poker hand!"

"Yes!"

It was shortly after 5:00 p.m. on Tuesday, November 9, 1965. We had never experienced a citywide power failure and assumed that a fuse in Le Figaro had blown. When the streetlights and traffic light at the corner of MacDougal and Bleeker streets also went out, we tried to comprehend what had happened, but still didn't. The only lights on the street were from passing cars. We walked back to Washington Square Park, thinking that surely the arch would be lit. From there we could see north to the tall buildings of midtown that were still reassuringly lit. But within minutes, as we stood with a gathering crowd, looking directly up Fifth Avenue, we could see the lights of the Empire State Building shut down, in bundles of floors, starting from the top. It was a disconcerting sight.

A queer calm engulfed the city's busy streets. Eight million strangers under a full moon watched skyscrapers all across Manhattan disappear one by one. Stories were being passed along about what had happened: The main power grid had had a breakdown, lightening had hit a hydroelectric power plant in Niagara Falls, the city was late in paying its electric bill, some lunatic at Consolidated Edison Co. had turned off the main switch. One story had an eight-year-old-boy run his electric train set into a puddle of water. It would take a federal commission to find out what had really happened on that night when thirty million people in eight northeastern states and Ontario lost all electric power.

Power didn't return until late into the night, when we heard stories about how magnificently New Yorkers coped and helped each other, about the eight hundred thousand people trapped in elevators and subways. I walked Uriah back to his apartment in Harlem, more than six miles uptown, meeting people along the way who told their stories. It was a friendly night in New York—so friendly that I had no fear of walking through Harlem with just the light of the silver moon.

Uriah and I had become good friends during our hours walk to his apartment. Three brothers, a sister, and his immigrant mother and father from Jamaica, all lived in one two-room, smoke-filled apartment. Here, the siblings had to fight for space at a dimly lit kitchen table to do homework amid loud neighbors' arguments frequently drowned, I guessed, by roars of commuter trains on tracks not fifty feet from the kitchen's grimy windows.

"This morning the professor said mathematics can tell us about the outcome of events; he said we could know if cigarette smoking causes cancer or if a drug can cure a disease. How's that, man?"

"Many things can cause cancer," I volunteered, and headed out for my long walk home.

A week later, we were back at Le Figaro. Our meetings were becoming routine. Uriah and I would go to the 14th Street Y and then walk to Le Figaro. The professor would be sitting in his usual window seat with his large empty cup of tea buried under crumpled sections of newspaper. We would invite ourselves to sit with him.

"It might seem that many cigarette smokers eventually get cancer," Uriah told the professor. "Man, if you take a sample of people from my neighborhood where lots of people smoke and drink themselves silly every night stressed over the money they win or lose at numbers, it may show that the number of people who get cancer is big—and, yet, smoking may not be the cause."

"The cells of the human body are constantly being bombarded by external elements and circumstances that are generally protected by the immune system," said the professor. "Cancer can be caused by any one of a number of things. You might think that we can't single out cigarette smoking, but we can. The first suspicion that there might be a connection came when women entered the work force during the war. The cancer rate for women approached that of men. Experiment design is tricky. If you're testing the effectiveness of a drug, you have to find out if it is truly the drug that cures and not the placebo."

"What's a placebo?" Uriah asked.

But the professor didn't answer.

"Why not just find out if cancer patients are smokers?" Uriah asked. "I mean, if a large number of cancer patients are smokers, then maybe cigarettes are not the only cause or not the main cause, but shouldn't we suspect them to contribute to cancer? Maybe the smelly black air coming out the back of buses contributes."[1]

The professor answered by way of what seemed to be a digression. He told a story about a lady tasting tea.[2] At an English tea party, one lady was overheard saying that she could tell by taste whether milk had been added to her cup before the tea or afterward. No doubt, that would take a finely discriminating palate. It would be harshly inhospitable to hold her precisely at her word, so her word was graciously interpreted as a more relaxed statement, suggesting that she might make some mistakes, but that most often she could distinguish whether milk had been added before or after the tea.

So an experiment was performed with eight cups of tea, four with milk added before the tea and four afterward. Clearly, if she were right with all eight cups, the experimenters would be convinced that she could discriminate. But what if she missed one? Would that contradict her word? Maybe not, but what if she missed two?

Uriah and I were puzzled. Not over the experiment, but over the connection to the question of how mathematics could determine the outcome of events. The professor's digression seemed to be more a matter of subjective judgment than an explanation of how mathematics could determine the outcome of events.

"Nothing is 100 percent certain in this world," the professor said. "So we must have a way of determining not what's certain, but rather what is probable. The lady should have permitted herself some possibility of error. After all, her taste buds would have changed after the first few sips; the milk, too, would have changed, waiting for her tasting. With such a delicate difference between tea poured before milk and tea poured after milk, it seems only fair to relax the notion of certainty and permit a few errors."[3]

"Remember the experiment you did in class with go chips?" Uriah asked. "You said it could be used to simulate election results and that it can just as well simulate weather conditions or a body's resistance to disease. I'm wondering how math comes into it."

"It's very simple to tell in theory, but difficult to do in practice. But I will tell you," the professor said putting his elbows on the table, again bringing his hands together so the fingers of his right hand could touch his left.

He went on to explain something that I should have understood from my classes in probability but didn't. A year of probability and statistics study was delivered in the time it took to drink a hot espresso.

"In the go chip experiment, we picked samples of twenty at a time. What if we were to think of each sample as picking twenty chips one at a time and replacing each chip before the next pick?" the professor said.

"That's not what we really did," Uriah interrupted.

"No, but we could think we picked the chips one at a time as long as we realize that there is a difference that will give us a slight error. The error comes about because scooping twenty chips depletes the bowl of some chips that would ordinarily be still in the bowl if we had picked only one at a time. We can make this error insignificant by increasing the number of chips in the bowl."

"You mean, if there were three thousand black chips and seven thousand white, the bowl will not miss twenty scooped-up chips?" Uriah asked while lifting his empty cappuccino cup, licking the sugar from its side.

"Right—the difference between picking chips one at a time with replacement would be very close to scooping up twenty at a time if there were ten thousand chips in the bowl with the same proportion of black to white. Now ask what the probability is in picking a single black chip. Because we know—even though we are not supposed to—that there are three thousand black chips and seven thousand white ones, we know that we have a three in ten chance of picking a black chip."

"So, the probability of pickin' a single black chip is 0.3, or three tenths, and the probability of picking a single white chip is 0.7," Uriah said, indicating he was following.

"Yes—so the question is, what's the probability of picking six black chips in a handful of twenty?" the professor asked.

"Why six?"

"Because six was the favored number in the samples of our experiment, remember?"

"Yeah, I remember."

"Well there are only two possibilities—a chip is either black or white. The probability of two such events happening is the product of the two probabilities.[4]

"Yeah."

"So the probability of picking k blacks and $(20 - k)$ whites from the bowl is $(0.3)^k(0.7)^{20-k}$."

"Yea, I get that. The probability of picking a single white is 0.7, *and* you want to pick k blacks and $(20 - k)$ whites. But I could pick three blacks in a row or pick a black on the first try and not pick another black until ten tries later, or lots of other possible combinations of picks."

Uriah could understand what the professor meant by k. It should have been too abstract for someone with almost no high school math background, but Uriah seemed to fully understand that k was standing for an arbitrary number—in this case, any number between 0 and 20.

"Right," the professor continued, "we must be careful because the same result can happen in a different order. There is a large number of possible ways of picking k blacks and $(20 - k)$ whites twenty at a time. That number is the *binomial coefficient* $C(20,k)$."[5]

"How do we know that?"

"I showed this in class. Maybe that passed by you, but trust me for a minute and I will come back to it."

"Okay," Uriah said. I could tell he was a bit embarrassed by the suggestion that he missed something that the professor had said in class, but it didn't slow him down. He kept interrupting the professor after every couple of sentences, as if the interruptions were helping him digest what the professor was saying.

"Remember," said the professor, "the probability of either one of two events happening is the sum of the probabilities.[6] So, the probability of picking k blacks and $(20 - k)$ whites in any order is, therefore, $C(20,k)0.3^k0.7^{20-k}$."

"Ah, the probability of pickin' k blacks and $(20 - k)$ whites in one specific order—say, three blacks in a row and the rest whites—is $(0.3)^k(0.7)^{20-k}$, but this could happen in either of $C(20,k)$ different ways, so your probability would be $(0.3)^k(0.7)^{20-k}$ added to itself $C(20,k)$ times."

"That's right."

"The probability of getting six blacks in any order would be $C(20,6)0.3^60.7^{20-6}$, which equals something near 0.192.

At this point, the professor jotted down a table of probabilities.[7]

Number of Black Chips in Sample	Probability of Picking k Black Chips from Twenty
0	0.001
1	0.007
2	0.028
3	0.072
4	0.130
5	0.179
6	0.192
7	0.164
8	0.114
9	0.071
10	0.031
11	0.012
12	0.004
13	0.001
14	0.000
15	0.000
16	0.000
17	0.000
18	0.000
19	0.000
20	0.000

The professor jotted down this table with almost no hesitation.[8] Did he make the calculations in his head, or did he memorize them? "It's clear that the highest probability is associated with $k = 6$," he said.

"So what?" Uriah asked.

"First, notice that $k = 6$ is favored. Then notice that you have almost a 95 percent probability that the number of black chips in your sample will be between three and nine. You can see that just by adding the probabilities associated with k from three to nine."

"Sure. Shouldn't we expect that?" I asked, just to say something.

"If we graph the probability against the number of black chips picked, we get something that looks like this." At that, he sketched the following.

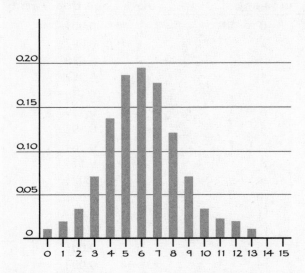

"This looks surprisingly like another curve that we know very well," he said and sketched a curve right on top of his last sketch.

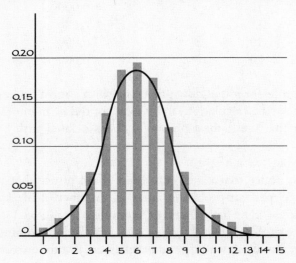

"We would have a better approximation by increasing the population size to, say, thirty-thousand black chips and seventy-thousand white ones. We can then construct a curve representing the probabilities that any one sample of size one hundred will come up with *k* black chips, where *k* can be anything from one to one hundred."

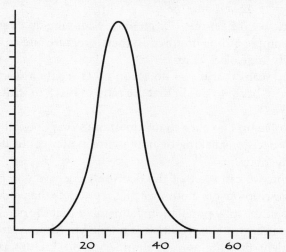

"The advantage of this last sketch is that we already know a great deal about this curve. It is completely determined by two numbers. The first is the *mean*, which equals thirty. (Remember, we are talking now about a sample size of one hundred.) It tells us the number of black chips favored in any sample size of one hundred—or, at least, it is the number that should be favored if the ratio of blacks to whites is truly thirty to seventy. The second number is called *the standard deviation*, which turns out to be 4.58, in this case."

"And what does the standard deviation tell us?"

"It tells us how concentrated the data is near the favored number.[9] But the magnificent thing about this curve is that the area under it tells us…."

"Hold on, hold on, what do you mean by how concentrated the data is?"

"I'm coming to that. The curve is adjusted so that the total area under the curve is 1. With that adjustment, we know how much area under the curve surrounding the favored number is 95 percent of the total area."[10]

"So, a pick that happens to fall outside of that area would be very rare."

"Yes, 95 percent of the area under the curve in question lies between $k = 21.02$ and 38.9768."

"Lemme see if I get this." Uriah said. "Scooping up twenty black chips in a sample of one hundred is, therefore, rare and so is scooping up thirty-nine black chips."

"Yes. It means that if you do scoop up fewer than twenty black chips, you can begin to doubt that the ratio of black to white chips is three to seven."

"So, professor, I go back to my question: What does all this have to do with decision-making or how mathematics could determine the outcome of events?"

"Ah, now we can see that the distribution curve one gets from sampling go chips favors a number that represents the percentage of black chips in the total population. Remember that the percentage of black chips in the total population is unknown. Also, we are 95 percent confident that the average number of black chips picked by repeated sampling from one hundred chips will be between twenty-one and thirty-nine. Picking fewer than twenty black chips or more than thirty-nine black chips is, we agreed, a rare event. Now suppose that you hypothesized that the proportion of blacks to whites was three to seven. Then, by your hypothesis, your distribution curve would look like the one in the figure above, with 95 percent of the area between twenty-one and thirty-nine. Right?"

"Right."

"Now suppose that you take a sample of one hundred chips and you find that only eighteen are black. Then shouldn't you doubt that there are really three thousand black chips in the bowl?"

"You mean I should doubt my hypothesis."

"Yes."

"And I should doubt it because...."

"Because if your hypothesis was true, it would be highly unlikely that you would have picked eighteen black chips in a sample of one hundred."

"But there'd be a 5 percent chance of getting eighteen."

"Yes."

"So what's wrong with 5 percent? Hey, I might make a numbers bet if I had a 5 percent chance of hitting a number."

"Yeah, but 5 percent?"

"If you think 5 percent chance is too large a room for error, you could make it 1 percent. It turns out that 99 percent of the area under the distribution curve is between eighteen and forty-two."

"So if you picked seventeen blacks or forty-three blacks, you have an even stronger reason to reject your hypothesis?"

"Right. In other words, the model built from the real-world statistical samples are not conforming to the true mathematical curve that models the hypothetical bowl with a three-to-seven ratio of black chips to white chips. It doesn't mean that there are not three thousand black chips in the bowl, so you might want to subject the bowl to further testing."

"I understand all this, but you still haven't answered my question: What does all this have to do with decision-making or how mathematics could determine the outcome of events?"

"What goes for chips in a bowl also goes for a host of other experiments. One could test the effectiveness of drugs: Is a certain drug 78 percent effective, as the drug company declares? Does a certain diet pill really reduce weight for 97 percent of the people who use it? Does a certain cancer drug really cure 24 percent of the patients who use it? Do 99 percent of the bombs deployed really hit military targets and not civilian targets? Does 80 percent of the voting public really support a particular congressional bill? Do 45 percent of all people who own TVs really watch the *Ed Sullivan Show* every week? Do 55 percent of Americans really approve of President Johnson's handling of the war in Vietnam? Do 40 percent of Americans believe a lone gunman shot President Kennedy? Is the Bureau of Labor Statistics correct in saying that the unemployment rate is 5.4 percent? Questioning a relatively small random sample of Americans can test all these truths."

"So, the experiments you'd perform to check on these truths are the same as the ones you'd perform to check on the ratio of black to white chips?"

"That's right. The chips can be people, and the random handful can be random picks of people from the population."

That was the year Lindsay was elected New York City's mayor; the year Central Park was closed to cars and open for biking, jogging, strolling; and the year that taste came back to the Big Apple. Tolerance of minorities was on the rise, but the Vietnam War was looming. Opinion polls were indicating ever-increasing unhappiness with Lyndon Johnson's war policies.

I didn't see or hear from Uriah or the professor after that, but I ran into Candice one day in a bookstore. The beauty of her eyes, which I hadn't noticed on that one occasion when she sat opposite me at the Broadway luncheonette counter, struck me as she told me that the professor had arranged a scholarship for Uriah and that he was matriculating in mathematics at Columbia.

CHAPTER 14

Boxcars and Snake Eyes

The Law of Large Numbers

To doubt everything or to believe everything are two equally convenient solutions; both dispense with the necessity of reflection.

—*Henri Poincaré,* Science and Hypothesis

omer tells us that gambling and chance have their roots in the beginning of time when Zeus, Poseidon, and Dis drew lots for shares in the universe.

Since we are three brothers born by Rehia to Kronos, Zeus, and I, and the third is Hades, lord of the dead men. All was divided among us three ways, each given his domain. I when the lots were shaken drew the gray sea to live in forever; Hades drew the lot of the mists and the darkness, and Zeus was allotted the wide sky, in the cloud and the bright air.[1]

Gambling might go back to the beginning of time when cavemen rolled stones the way children now flip baseball cards. Unlike other mathematical theories—of numbers, geometry, and astronomy—the theory of gambling had, more or less, a definite beginning in the year 1654.

The winter of 1654 was unusually cold for Paris. Even the Seine froze. Hundreds of people skated and slid on the river while fires burned at street corners and parish priests distributed bread to the poor. Those who had money suppressed their spending to provide food and warmth for the poor people who could not find work that difficult winter. Thirty years of religious wars in Europe drained the treasury of France. With huge debts and expenditures accumulating at alarming annual rates, France was forced to increase taxes on its working classes. But dishonest tax collectors and agents brought very little revenue back to the treasury. Exempt from taxation, the nobility continued to accumulate appalling excesses of wealth. This was during the early reign of Louis XIV when France was becoming the leader of Europe, a time when the idle rich were overtly gambling in gambling rooms all over Paris.

Early one cold winter afternoon in 1654, the sun broke through the swiftly moving clouds passing over Rue des Pastes as Antoine Gombauld stepped onto a hackney coach to take him to one of his favorite gambling rooms on his way to court. A distinguished nobleman and gambler, nicknamed Chevalier de Méré, knew an old gambling rule that should have given him good odds on betting that he could throw at least one double six with twenty-four throws of a pair of dice. He thought he had the mathematics of gambling worked out. As his hackney lurched forward, he must have wondered if there could be some mathematical explanation. For the remainder of his thirty-minute ride, a probing mind like his could have contemplated the idea of using mathematics to his gambling advantage.

De Méré, before his appointment at court, could have ordered his driver to bring him to the bookshop of his friend Pierre Froullé on the Quai des Grande Augustins, a short walk across the Seine from the palace. The coach would have turned right onto Rue de la Harpe (now Boulevard Saint-Michel, the "Boul' Mich.") one of the better gravel roads leading to the Pont Saint-Michel. De Méré was in the habit of browsing the bookshops between the Pont Saint-Michel and the Pont Neuf amid invigorating smells of roasted chestnuts, ciders, and vinegars wafting from Maille's, a nearby vinegar distillery. The old bookseller, Mr. Froullé, collected the finest editions of classics and sometimes reserved newly acquired works for his special bibliomaniac clients. Upon seeing de Méré walk into his shop that morning, Mr. Froullé, who would have known Gombauld's reputation as a gambler and intellectual, brought out a large envelope marked *Liber de Ludo Aleae* (Book of Dice Games). It was a hundred-year-old unpublished Latin manuscript, filled with important contributions to calculating probabilities connected with gambling by Girolamo Cardano, a Milanese physician, mathematician, and gambler, better known for his 1545 published book *Ars Magna* (The Great Art), an account of everything known about the theory of algebraic equations up to the time. De Méré bought the manuscript and most likely browsed it before walking on to the palace. His fortune was quite considerable, but he was beginning to lose heavily in dice at the gambling tables and was wondering why the old gambling rules were no longer working in his favor. Along the Quai des Grande Augustins he met his friend Pierre de Carcavi, who was watching Parisians skate on the Seine. Despite the cold weather, the hurdy-gurdy players, organ grinders, fire-eaters, letter writers, and sundry street performers were surely in their usual corners near the palace. Paris was still a small medieval walled town in 1654, densely filled with half-timbered houses and stone government buildings, encircled by thirty-foot walls and moats with gates, and sentry towers with trees growing through their centers. De Méré and de Carcavi would have walked together and perhaps stopping near a warm street fire, de Méré confessed his growing gambling addiction and the huge toll it had taken on his personal fortune while de Carcavi read a few handwritten pages of the *Liber de Ludo Aleae*.[2]

De Carcavi would have translated aloud and relayed the dice problem to de Méré. Amazed that Cardano had already had rudimentary ideas of probability theory a century earlier, de Carcavi began to work out the theory for himself, following the premise that rarity of events can be mathematically justified, not simply consigned to luck or chance. He quickly realized the significance of the problem and, as was his custom, sought the advice of his good friend Blaise Pascal.

Two questions influenced the manuscript. One asked for the

smallest number of times a person must throw a pair of dice to have a better than even chance of getting a double six. The other, known as the *problem of points*, asked for the number of points that should be awarded to each of two players in a game of dice if the game is left unfinished. This was precisely the information de Méré needed to reverse his luck.

Pascal spent the next few days examining the manuscript for signs of a solution to the dice problem, but he was skeptical of its results and wrote to his friend Pierre Fermat about the problem. Pascal became ill and, confined to his bed during the spring and summer, corresponded with Fermat and worked to extend the problem of points to more than two players.

Pascal felt that in a fair interrupted game of dice, each player should be awarded points according to the probability of their potential in winning. He worked out the mathematics of the prizes and proved that the odds are slightly less than even that double six would turn up on twenty-four throws of a pair of dice and slightly more than even that double six would turn up on twenty-five throws.

How can the theoretical mathematics of an idealized pair of dice predict the behavior of real dice thrown by a real person? After all, the dice are imperfect white cubes with rounded edges, presumably made in such a way that the indented black dots do not disturb their rotational symmetries. Presumably, the manufacturer has accounted for the fact that the six slight gouges that make the six black dots do not deplete enough material to tilt the cube toward the side with one slight gouge.[3] Predicting what a real pair of dice will do when thrown requires a new kind of reasoning scheme, a logic more closely related to inductive reasoning than deductive reasoning. Typical mathematical reasoning is based on a collection of unarguable, self-evident assumptions, together with cognitively unshakable rules of deductive logic. Truths are derived from chains of truths linking, often invisibly, back to the self-evident assumptions; in other words, we find truth from truths we already know. We might even develop absolute truths to predict the behavior of an ideal pair of dice. But will those truths have any direct relationship to those that tell us about real dice?

Euclid talked about points and lines as if they had no thickness. A point, to him, was an abstract location; a line was self-defined by the self-evident truths designed to develop geometry. What does Euclid's line have to do with the line drawn by the sharpest pencil in the world, scored along the straightest straight edge in the world? Almost nothing, whether Euclid's line is thickly drawn by clumsy human hands or by machines attempting to approximate it. And yet Euclid's geometry seems to work well as a model for many real-world applications, as long as measurements are made finely enough. In other words, idealized mathematical models can be useful when they fit closely to the real-world phenomena they are designed to explain. But how close is close enough? We might know that the probability of rolling "snake eyes" with an idealized pair of dice is 1/36. But does that mean that I should expect to roll snake eyes once in 36 tries? No.

So what should we expect to happen when we humans roll a real pair of dice 36 times? To answer this, let's ask a simpler question. How many times would we need to flip a penny for heads to come up once? If the coin is fair, the probability of heads is 1/2—after all, there are only two sides to a coin, and heads is only one of those sides. If tails came up on the first flip, should we expect heads to come up on the second? No, of course not. But there is a strange law called *The Law of Large Numbers* that tells us that if we flip a coin long enough, half the time it will come up heads and half the time it will come up tails. Suppose that 75 tails and 25 heads are the result of a hundred flips. Does that mean that for the next hundred flips you will see more heads than tails? Of course not. After all, how could the penny know how many tails and heads appeared before?[4] We have a very odd kind of law—odd because, on the one hand, it suggests that it is a law and, on the other, the penny keeps a record of its history. The law really says that after flipping a large number of times, there will be a tendency for the number of heads to be close to the number of tails. The law does not say anything about how large the number of flips must be; it only says that as the number of flips grows large, the number of heads will grow closer to the number of tails.[5]

What does The Law of Large Numbers say about rolling snake eyes with a pair of dice? It doesn't say that it will take thirty-six rolls to get one snake eyes.[6] It says that you are likely to roll snake eyes once if you roll it thirty-six times, but that you are much more likely to roll snake eyes one hundred times if you rolled the dice thirty-six hundred times. And, of course, you are even more likely to roll snake eyes a million times if you rolled the dice thirty-six million times.

It is a nasty law, for its misinterpretation is responsible for a great many gambling mistakes. Presuming that the event remembers that its history is an infamous gamblers' mistake, called the *Monte Carlo fallacy*, it whispers into the novice gambler's ear at roulette wheels of casinos everywhere. "After a long run of red, bet on black," it says. There is something compelling about placing a bet on black after a long run of reds. Placing your bet on 0 puts your chances of winning at a probability of 1/37, so you are more likely to win by choosing one of only two possible colors rather than one of thirty-seven numbers. Would you bet on 0, if 0 did not come up in the last hundred spins? The probability that it would come up on the hundred and first spin is exactly the same as it was for any of the first hundred spins.

So how can we say The Law of Large Numbers is a law if it doesn't seem to obey any predictable behavior? When Uriah posed this question to the professor one afternoon at Le Figaro, the professor had us walk to a nearby candy store to buy five small boxes of Indian Brand sunflower seeds and a box of chalk. On the sidewalk just outside the store, he paced off a large square three feet wide, judged by the length of his shoe, and used the chalk to draw the square. Centered inside the large square, he marked off a smaller square one foot wide.

"Watch this," he said, throwing the contents of the first box of sunflower seeds into the air directly above the squares. Most seeds landed inside the large square, but a few bounced outside both squares. "Here," he said handing each of us a box. "Throw the seeds into the air right above the large square."

We threw the contents of our boxes into the air and watched the seeds fall inside the chalk squares.

"Now, just for good luck, I will throw the seeds from the remaining two boxes into the air," he said while opening the boxes, throwing the contents into the air and asking, "What should the ratio of seeds in the small square to seeds in the large square be?"

"Nine to one," Uriah quickly answered.

"Right," said the professor. "The probability of a randomly thrown seed falling within the smaller square is 1/9. Do you see why?"

"Of course," said Uriah. "The Law of Large Numbers says that if you completely covered the large square with sunflower seeds, you would have the area in square sunflower seed units. But you would have also covered the small square and, therefore, also have the area of the small square in square sunflower seed units."

"Exactly!" shouted the professor.

We counted 128 seeds in the big square and 14 in the small square. Though odds slightly better than nine to one should not have surprised us, we were stunned to find the theory in such close support of reality.

The Law of Large Numbers influences the distribution of random events. To experience this, you could photocopy the illustration on the next page at a high magnification onto an 11×17 sheet of paper and take a handful of peppercorns—say, 100—and let them fall randomly on the picture. Count the number of peppercorns that fall within the dark shaded area, and divide by the number of peppercorns that fall under the curve and above the straight line at the bottom. You will find the result to be close to 0.68, which means that approximately 68 percent of them will fall under the curve within the middle shaded area. If you divide the number of peppercorns that fall within all shaded areas by the number that falls under the whole curve, you should get a number close to 0.95, which means that approximately 95 percent of them will fall under the curve within the shaded areas.

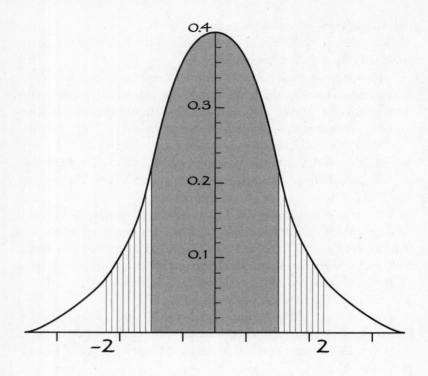

To appreciate why your random scattering of peppercorns distributed according to such percentages, try the following experiment. Buy a large box of Good n' Plenty candy. Black and white go stones would be better, but Good n' Plenty will do. The box contains candies that look like an assortment of pink and white pill capsules. Mix the candies in a large bowl, and pick out ten candies at a time without looking. Count the number of pink candies; label that number P_1 and call it a pink distribution number. Replace the candies, mix the bowl, and pick again to get another number, P_2. Repeat this procedure another twenty-eight times to get numbers P_1, P_2, P_3, … P_{30}. (The subscripted numbers give labels to each pick.) Now make a graph. Draw a horizontal line with ten equally spaced intervals labeled zero through ten. Count the number of Ps equal to 0, and call that number D_0; count the number of Ps equal to 1, and call that number D_1; and so on, until you get to D_{10}. Then graph all eleven of the D_i as a bar chart. The figure below is an example of what typically happens.

I have performed this experiment many times as a class demonstration to show how to learn something from organizing random samples. Though I know I shouldn't be surprised, every time I perform this experiment, I am astonished. The graph below could also represent the distribution of black go stones randomly scooped, ten at a time, from a bowl of black and white stones. If there were one hundred stones in the bowl, there should have been about thirty black stones.

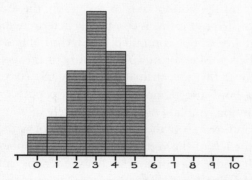

An increase in the number of samples—say, from thirty to three thousand—would give a bar graph that closely follows the curve below.

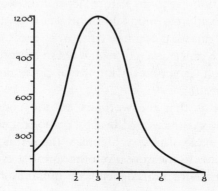

The shape of this curve is known and can be computed simply by knowing one constant, the total number of black stones in the bowl. This appears to do us no good because we are not supposed to know the number of black stones in the bowl. But argue backward and see what happens. In this particular case, the total number of black stones is thirty and the total number of stones is one hundred. This means that the likelihood of picking a single black stone from a pile of one hundred is 30/100, or 0.3. This is all we need to construct the curve above. The curve shows that 68 percent of the time, you will pick more than one stone and fewer than five. Ninety-five percent of the time, you will pick more than zero and fewer than six stones.

It's all in the shape of the curve. Draw a vertical line through the real number x on the bottom axis of the graph, the area under the curve and to the left of the line is completely determined by the curve itself, and that area can be computed. The technicality of computing the area can be set aside for now. The beauty of this is that we have related two things that are, at first, not clearly related: picking black stones from a bowl and randomly scattering peppercorns over the earlier distribution graph. The area to the left of the vertical line through x is approximately the ratio of peppercorns that fall under the curve to the left of the vertical line over the number of peppercorns that fall under the curve. The area under the entire curve is equal to 1 by design, so as the two vertical lines move outward, it becomes more certain that the peppercorns will fall between the lines.

It is almost always true that experiments in science involve measurements, and that events in nature depend on so many variables that exact observation is impossible. If exact observation is not possible, then how can we determine truth in science? Cigarette smoking might lead to cancer, even though it doesn't directly cause cancer. Does the Surgeon General really know that cigarette smoking might be harmful to health?

The answer is that the world is far too complicated to be explained by laws or measured by observations. Only 4 percent of the universe is made of observable mass; the rest is anywhere from hard to impossible to measure. Even the simplest event, such as the flip of a coin or toss of a die, depends on millions, if not billions, of undetectable happenings, so one can only imagine how many undetectable happenings are responsible for a complex phenomenon such as cancer. Statistics provide us with a measurement of the likelihood of whether people who smoke will get cancer. However, this measurement doesn't actually say anything about a cause-and-effect relationship, such as *why* people who smoke get cancer. In general, many natural relations that cannot be explained by laws or measured by observations can be linked by statistical measurements. In many experiments, probability is a measure of area under a curve that describes how events distribute themselves.

Is it possible that there is a pinch of randomness included in the recipe of the universe?

> "First of all, the Chaos came into being.... Out of Chaos came darkness and black Night, and out of Night came Light and Day, her children conceived after union in love with Darkness...." (These are the beginning lines of Hesiod's *Theogony*, an eighth-century B.C. history of the divine government of the universe.)[7]

When science discovers law, it does so by discovering a pattern in what appears to be randomness. Science sees the light and day from the darkness and night. The vastness of the universe and the relative infancy of science explain the overwhelming imbalance between randomness and law.

Randomness has become a positive feature of science. Many laws of science now embody randomness as a necessity. Brownian motion (the curiously random movement of microscopic particles suspended in gases or liquids caused by the constant bombardment of gas of liquid molecules) is not only used as a model of the randomness of the path of a typical molecule in a gas or liquid, but it also is used to make predictions about the average behavior of many phenomena that are governed by randomness. Even if we cannot define an exact pattern in the events that we investigate, we might be able to place some measurement on the difference between what we see and what pattern (or law) we expect our observations to follow. This measurement is precisely what the science of statistics is about. Modern statistics began to take root in the late nineteenth century with the question of how to deal with events that are associated with sets of possible outcomes, where any one outcome is thought of as a value of a random variable. That random variable is expected to distribute itself over a range of possibilities. There are clairvoyants who profess to see things that are beyond ordinary perception. Why not, if only 4 percent of the universe is made of observable mass and much of the rest is immeasurable? A clairvoyant might claim that he or she could foretell the sex of unborn children—without ultrasound. How would we be able to distinguish between random guesses and true perception? After all, the sex of the unborn child is determined randomly, and so is the guess. The right distribution curve depends on the design of a good experiment. The area under it plays the same role in determining the truthfulness of the clairvoyant as it does in determining where a peppercorn will fall.

Today the probability measurements are tabulated so that it is no longer necessary to compute integrals or find areas under curves. However, without the mechanics of calculus, these integrals could never have been evaluated and science would never have come to Light from Chaos, darkness, and black Night.

In Part 1 we distinguished between the logical validity and truth of an inferential argument while developing a background for deduction. We were also concerned with the distinction between the analysis (reasoning backward to established facts or principles) and synthesis (reasoning forward from established facts or principles) of mathematics, hoping to suggest that mathematics can advance more by intuition than by formal concerns of logic. Here, let's explore the idea that inferential arguments might be considered not for their absolute certainty, but for their likelihood. We call this idea inductive reasoning, subjective reasoning, or plausible inference, an idea suggested back in the seventeenth century by Francis Bacon.[8] Though a statement such as Carl Hempel's humorous argument (described in Chapter 11) that "all swans are black" might not be true, we could try to attach some weight to what it means to be a swan and what it means to be black. To show that all swans are black, we applied an inductive argument. An inductive argument runs something like this: To be convinced of a general truth, you only need to have many specific experiences of the truth.

Take the example suggested by David Hume[9] and examined by the economist John Maynard Keynes[10]: "Nothing so like as eggs." It's a short phrase, not an argument. But it conjures up so much past experience with eating eggs that there is little that we can do to avoid the subjective feelings about eggs that comes from the many times we have eaten, seen, or just smelled them. If we are to now put that phrase into an inferential statement such as "All eggs are good," we could easily accept not only the validity of the statement, but also its truthfulness. And yet we have not had the experience of eating *all* eggs. Keynes suggests that Hume "should have tried eggs in the town and in the country, in January and in June," or, as Dr. Seuss suggested, "In a box, with a fox, in a house, with a mouse." Keynes goes on to say that Hume "might have discovered eggs could be good or bad, however like they looked." Still, Hume could have insisted, "I do not like them here or there. I do not like them anywhere!"

The question is how we're persuaded by multiple instances of an event. Is it better to have the experience of a large number of identical events or by a smaller number of varied events? If eggs are good, we should be able to establish their goodness in variety. Otherwise, we would have to qualify what we mean by eggs.

For another example, the flow of electrons in a wire is called an *electric current*. The simplest circuit is a wire attached to a battery. This sets up an electric current, which is defined to be the amount of electric charge that flows through a cross-section of the wire per unit time. Charge is one of the fundamental measures of matter. Put an open switch in the circuit, and the electrons still flow, but in a random direction as they collide with the fixed atoms that constitute the wire. There is no net direction to the flow of electrons. But when the switch is closed, there is a net flow of electrons from the negative electrode to the positive. This flow is the electric current; it can be measured with a device called an ammeter.

Even when the switch is closed, the current is not precisely determined because the random collisions of the electrons with the stationary atoms interfere with the flow. Of course, the ammeter might not be accurate enough to pick up any differences between current flow from one second to another, but there are differences. Ohm's law says that the voltage V is the product of the current I and resistance R: $V = IR$. This means that the voltage is completely determined by the current and resistance. That would suggest that Ohm's law is a deterministic model for the relationship between the measurements of voltage, current, and resistance. If you check the voltage with a voltmeter at several different times, you will find that it has not changed. But there must be deviations. We could say that the deviations are so small that we want to ignore them. It would be our decision to think of Ohm's law as being deterministic.

In contrast to Ohm's law, we can look at the meteorological question of how to predict precipitation from other measurable variables, such as wind velocity, barometric pressure, and pressure changes. No deterministic models relate the relevant variables, so we are forced to explore probabilistic models.

Take another example. Suppose that I have a tank filled with water and inject a small line of a dissolvable substance such as finely ground cereal (Postum, Inka, or permanganate) on the left (see illustration below). In the illustration, the dots indicate the substance as its concentration diminishes from left to right. Leave the system still, and you will notice that the substance spreads from left to right, from higher concentration to lower, until it is equally distributed through the water.

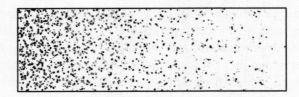

Now you might think that there is some force that is driving this tendency of molecules to move from the more crowded region to the less crowded one. But there is no such force. There is no more a force that drives this than there is a force for evolution to drive organisms to become fitter. Every molecule in this system is independent of all the other molecules in the system. Every molecule is being knocked about by impact with the water molecules and, thereby, being knocked about in an entirely unpredictable direction. The path of any dissolved molecule is determined randomly. To understand the truth of what is happening, place an imaginary wall across the tank and ask how probable it is that a molecule on the imaginary wall will move toward the right. The answer is that it is equally likely that it will move to the right as to the left. If you take this to be true, you must agree that more molecules will move from left to right than from right to left, simply because more molecules are coming from the left side of the imaginary wall than from the right. So, the diffusion toward uniformity occurs simply because there is an equal likelihood that the molecules will move in any direction.

The second law of thermodynamics tells us that you can play the same game with gasses. Take two containers, one with gas at some pressure and the other empty. Connect the two by a tube that lets the gas move freely between them. The gas will quickly spread until both containers have half the starting pressure. This equalization of pressure is one example of a universal tendency of particles to distribute themselves in as many ways as possible. We measure this tendency with a variable called *entropy*. Here is the surprise: The gas molecules will randomly bounce off each other like bubbles in a pot of boiling water so that, over time, each will find itself, for a while, back in the container it started from. Henri Poincaré demonstrated this in a general theorem about dynamical systems.

Imagine two containers, labeled A and B, and a bucket load of chips numbered from one to one hundred. In container A, one hundred balls are placed, each labeled uniquely with numbers from one to one hundred. Container B starts out empty. A chip is picked at random from the bucket and its number (N) read. The ball numbered N is moved from container A to container B. The chip is replaced, the process repeated, and each time a random chip is picked. What will happen? Poincaré's general theorem predicts that the number of balls in container A will decrease at an exponential rate until both containers have approximately the same number of balls. As the number of balls in container A decreases, the likelihood of picking a chip marked with a number from A decreases. The rate of decrease is proportional to the number of balls remaining in container A; that's why the rate is exponential. But here is the surprise: With absolute certainty, all the balls will eventually return to container A, although it might take an enormously long time for that to happen.[11]

Sir James Jeans, the late renowned physicist who was knighted for his contributions to astronomy and the popularization of physics, used to quip that anyone still breathing today is breathing in the molecules of the dying breaths of Julius Caesar.

Imagine what would happen if you placed a large number of fleas at the center of a checkerboard. Very quickly, the fleas would begin to jump in all directions to fill the checkerboard. But the fleas are simply jumping around without any predetermined direction. Any one flea is not jumping with a space objective in mind—it does not know in which direction to jump to have more space to itself, for even if it has lots of space, it will jump again in a new random direction.

Chance makes the world go around. From randomness, it creates an ever-evolving dynamic world in which electrons accidentally collide in a copper wire; permanganate diffuses with no purpose yet homogenizes in water; gas shares pressure with a vacuum, satisfying laws of thermodynamics, yet getting a chance with time to reverse its behavior; fleas jump aimlessly yet spread out on checkerboards; and DNA erroneously replicates itself without a plan, and, so, fortuitously, creates humans of varying ability.

CHAPTER 15

Anna's Accusation

Tests for Truth

A meaningful physical discussion always requires an operational background. Either this is provided by an existing theory, or you have to give it yourself by the sufficiently explicit description of an experiment that can, at least in principle, be performed.

—*David Ruelle*, Chance and Chaos

Stand at the northwest corner of Bleecker and MacDougal streets in New York City, look southeastward, and what do you see? Le Figaro is there just as it always has been. It is now a café, not a coffeehouse. Now, there is hardly a café in the city that doesn't sell good cappuccino. But my clever professor can no longer be found. You might ask his name, but I never knew it. We referred to him as "the professor." When we addressed him, it was "Professor." He is now a memory, a Mr. Unis, a man who could stand on one finger, a memory changed by distortion and interpretation of facts, like a memory of the shape of a room or the color of a car. When we asked about him, he told us more about his father than about himself.

His father's name, as I recall, was Edmond, a Belgian who was a young man late in the nineteenth or early twentieth century.

After his involvement in a terrible accident, Edmond visited a fortuneteller, who told him that he would fall in love with a beautiful rich young woman, and that the father of the young woman would disapprove of the romance and would concoct a plan to thwart it. The fortuneteller went on to say that through the forces of circumstance, the two lovers would meet again, but only after much time went by. Edmond was the poor sixteen-year-old son of an Antwerp fisherman, who, with his father, kept a small boat at a little stone jetty on the far other side of the Antwerp harbor. Looking south, across the great bay, from the wobbling little boat and the stable stone jetty, Edmond could see the great tall ships anchored on the south side of the harbor. "I'm proud to live and work in this great harbor," he entered in his diary one day. "I feel a part of the big unknown world," he entered on another.

One morning a beautiful, flirtatious young woman was standing behind a large camera just beside his boat. Her name was Eugenia. Smiles were exchanged, and Edmond fell in love with her. She was the daughter of a nobleman, the Marquis de Mollineux, so there was no chance for the encounter to amount to anything more than a flirtation. Coincidentally, the marquis accepted the responsibility of becoming Ambassador to the Vatican and, in so doing, made plans to move his family to Rome for the following year. The *Star of India* was to take the marquis and his family to Rome. Edmond boarded the *Lotus* in search of Eugenia, but before he discovered that he was on the wrong ship, the *Lotus*, laden with cargo for the Congo, left the port of Antwerp and headed for the open seas. A week later, the *Lotus* docked at Matadi, the port town at the mouth of the Congo River.

What does a young Belgian fisherman do in the Belgian Congo at the turn of the twentieth century? The only choice was to work for ivory tradesmen or one of King Leopold's rubber companies. For six months, he worked for the King's Anglo-Belgian India Rubber and Exploration Company as an overseer. That meant canoeing up the Congo River from village to village with a platoon of soldiers seizing women and children as hostages until the chief of the village would agree to send his men into the jungle to bring back specified amounts of rubber. When the proper amount of rubber was gathered, the women were sold back for chickens and goats. After six months of this darkly cruel work, he heard that a diamond mine was about to open in Luebo, near the Kasia River.

Edmond spent the next two years in Luebo, earning only ten francs a week. The marquis's daughter forgotten, his life became rooted in the mines, though he planned to leave just as soon as he could save enough. But save as he tried, he could not save enough from the few centimes that he earned from his hard labor. He spent a good part of the second year on a plan to steal a diamond, only to conclude that theft would have been too risky. Every miner was thoroughly searched upon leaving the mine. Life savings changed hands late at night when men played craps.[1]

Not having any organized betting scheme, his luck was strangely good. Centimes turned to francs, and before he knew it, he had more than enough money for passage home.

On the voyage home, without any earlier hints of mathematical knowledge, he discovered some mathematical secrets of successful betting. He realized that snake eyes and boxcars (double sixes) very rarely turn up because they have a 1-in-36 chance of turning up, whereas seven has a one-in-six chance. With no knowledge of Chevalier de Méré or Pascal, Edmond attempted to explain the odds of betting on various dice combinations and calculated the odds of throwing one double six in four throws. He found the probability of not getting any sixes in four throws. (This is the correct way to tackle the problem.) "If the probability of throwing a double six on a single throw is $1/36$," he thought, "then the probability of *not* throwing a double six on a single throw must be $1 - 1/36$, or $35/36$." Because each throw is independent of the previous throw, the chance of not throwing a double six twice is just $35/36 \times 35/36$. The chances of not throwing a double six three times are $35/36 \times 35/36 \times 35/36$. And the chances of not throwing a double six n times are $(35/36)^n$. So, the probability of throwing a double six n times is $1 - (35/36)^n$. "Aha," said Edmond. "What is the smallest n for which $1 - (35/36)^n > 1/2$?" Edmond was asking the same question Chevalier de Méré asked in the winter of 1654: What is the smallest number of throws that will give a better than even chance of throwing a pair of sixes?

"The solution is correctly marked in the diary," said the professor. "Edmond solved the equation $1 - (35/36)^n = 1/2$ and gets n to be 25, to find what de Méré knew more than three centuries earlier:[2] that if you bet you could throw a double six in 24 throws, the odds are against you, but if you bet you could throw a double six in 25 throws, the odds are in your favor."[3]

Old Monte Carlo whispered its familiar fallacy to Edmond, who, with more gamblers' blood than a fisherman's, was lured to frequenting Antwerp gambling halls by his incessant good luck. His growing wealth and lifestyle connected him to circles of the rich and, inevitably, to Eugenia. To impress her, his gambling frenzied to such persistence that he computed mathematical odds in his dreams. But gambling tables follow funny laws: Like light that is both a particle and a wave, they are strikingly mathematical and conspicuously chaotic. He began to lose. The more he bet, the heavier his losses were. Before long, he was as poor as he ever was.

The Monte Carlo fallacy is only one of many false intuitions adopted in gambling. Americans spend billions of dollars on lotteries every year, guided by a mistaken belief that comes from three misguided rules: one, someone has to win; two, I have as good a chance as anyone else; and, three, I really think my chances are pretty good. Plausible cause is twisted into feelings of truth. Better than even odds evoke the words "most of the time." *Most* does not mean 100 percent. Nor does it mean more than 60 percent. "Most of the time" simply means a majority of the time, which means more than 50 percent of the time. It's a useful exaggeration tool, used either when we really don't know the facts or when we purposely want to mislead someone into imagining a percentage much closer to 99 percent than to 51 percent.

Consider the following argument:

Most convicted felons are guilty of the crimes they committed.
Calvin Johnson Jr. is a convicted felon.
Therefore, Calvin Johnson Jr. is guilty of the crime he committed.

Calvin Johnson Jr., a black man who spent sixteen years in prison where he was serving a life sentence for rape of a white woman and burglary, was exonerated in 1999 by DNA evidence. DNA "fingerprinting" is relatively new to forensic science. Before DNA testing, blood typing, serology, and conventional fingerprinting were the standard tools. These conventional forensic tools give very imprecise measurements compared with DNA fingerprinting. About 40 percent of Americans share type O blood, and matching fingerprints are inconclusive in many criminal cases. Peter Neufeld, a pioneer in forensic psychology, called DNA indentification the "gold standard of innocence." Barry Scheck, one of the attorneys on O.J. Simpson's defense team said that DNA identification is the "magical black box that suddenly produces the truth."⁴ DNA fingerprinting is now playing a major role in exonerating wrongly convicted prisoners. Still, defense or prosecution lawyers could play DNA testing to their advantage, either by impressing the jury with its unassailable scientific accuracy or by attacking the evidence collection and storage procedures. In the O. J. Simpson case, the prosecution had substantial DNA evidence, but the defense was able to persuade the jury that the evidence had been tampered with.

The earlier syllogism is not the kind of argument that mathematicians would accept, but there is a place for it. It is an argument subject to a different kind of logical inference. We might call it probable inference. What do we mean by "most"? What number of guilty convicted felons would be satisfactory? The Truth in Justice Organization reports hundreds of wrongful convictions and claims that "eyewitness identification is one of the most potent and effective tools available to police and prosecutors. It is compelling, and time after time, it convinces juries of the guilt of a defendant. The problem is, eyewitness identifications are WRONG at least 50 percent of the time!"

For another consideration, I offer the true story of Anna, an autistic child whose father was tried, convicted, and sentenced to seven years in prison for having sexually molested her when she was a toddler. The jury was persuaded that Anna's father was guilty through testimony that involved indirect messages from the girl to a communication facilitator.

Anna was an autistic high school student when facilitative communication (FC) was in vogue as a method to help cerebral palsy children communicate. It allegedly works by having a "trained facilitator" support a communicator's finger and palm while he or she spells out words by selecting letters on a letter board, typewriter, or computer. In theory, it is assumed that the words are coming from the person with the disability, presumably something that he or she wishes to independently communicate.

One day, Anna spelled out a message charging that her father had sexually molested her when she was a toddler. A district attorney brought her father to trial and, using an indirect message that might have come from the girl or the facilitator, convinced the jury that the father was guilty. Anna's father was sentenced to seven years in prison.

Should we believe in FC or be skeptical? If the messages were coming from the communicator, ignoring or banning FC would deprive persons in need of an important part of life and culture. If the messages are coming from the facilitator and not the communicator, those messages can be dangerous. Some messages have accused innocent persons of sexual abuse and other criminal behavior. Moreover, if FC is ineffective, it can raise false hopes and take away scarce resources from other methods or activities that could benefit persons in need.

The problem for us is to distinguish between belief and reason. How can we reconcile the difference between what we believe and what we have concluded through deduction from undeniable certainties? To complicate matters, it is easy to fall prey to false beliefs, or prejudices, even when we seem to be carefully aware of our reasoning procedures.

A six-year-old girl with autism communicated the following message:

> i am really the only child in my class who uses a typewriter. i feel very proud of my typing. i will be a writer when i am 25 years old i x [sic] want people to respect children with autism. we are bright and want to be just like opther [sic] kids.

There are only two typographical errors (not spelling mistakes) in a relatively complex, content-filled paragraph. It would be an example of advanced writing even if it came from a child of the same age with normal communication ability. The message by this one six-year old girl surely does not suggest that FC is a hoax. It does suggest that, in this case, the message came from the facilitator. But there are many other examples of communicators typing relatively complex messages.

We should not be persuaded one way or another by one message. There are other tests. Several have been performed to scientifically determine the validity of FC. Try typing a short message using just one finger held in the air above the keyboard without looking at the keyboard. Even a skilled typist cannot complete a sentence without making an extraordinary number of mistakes. And yet the communicator does not look at the keyboard, whereas the facilitators do. Doesn't this suggest that it is the facilitator who is doing the communicating, not the communicator? I put a skilled secretary to the test. I asked her to type a short sentence using just one finger held in the air above the keyboard, without looking at the keyboard, just as a trained facilitator would do. She typed the following:

Kjrp[ikpjiiop gkf koiirf ogr ;,. dfsk j;l ijom ijm kikiu.

Translation: The quick brown fox jumped over the cat's back and then ran away.

Still, this should not persuade against FC. Perhaps in time, one could learn to type without looking at the keyboard with just one finger held in the air. We need a more plausible test.

Suppose that we perform the following admittedly oversimplified experiment.[5] Call the child Anna and the facilitator Barbara. Anna and Barbara are shown two pictures at the same time, say, of familiar household objects or animals. The pictures each person sees could be the same or different, chosen in a double-blind way (that is, chosen so that the investigator does not see the pictures and, therefore, is not subconsciously influenced by unneeded knowledge), but neither Anna nor Barbara is permitted to see what each other sees. Anna, with the aid of her facilitator, types a description of the picture she sees. Hypothetically, what gets typed is a description of the picture seen by Anna. It's easy to determine whether Barbara is facilitating by simply not letting her see any picture and then marking how often Anna writes down what she is seeing. But to show that Barbara is actually imposing her own thoughts onto Anna's communication, we need more. We need to see how often Anna types what Barbara is seeing. The statistician might suggest a notation such as (x,y,z) to represent the three objects as follows:

x = the pictures Anna sees
y = the pictures Barbara sees
z = the pictures identified by the typing

After being shown a great many pictures, the experimenter should have two central questions:
  1. What percentage of triples (x,y,z) have $x = z$?
  2. What percentage of triples (x,y,z) have $y = z$?
But what should those percentages be? The experimenter should not expect the answers to either question to be 100 percent. What percentages of the answers would begin to persuade us that the typed response is coming from Anna and not from the facilitator? What percentage of answers would convince us that a facilitator does make a difference?

The statistical analysis of such an experiment is too complicated to be discussed here. Such double-blind experiments have been performed, resulting in significant reasons for rejecting the idea of facilitated communication. In study after study by independent investigators, the pattern appears to favor these opinions: The communication comes from the facilitator, not the child, the facilitators are not aware of their influence on the communicator, and communicators fail almost all objective testing.

The O. D. Heck Developmental Center conducted hundreds of trials involving twelve students and nine facilitators. "There was not one single correct response. There was overwhelming evidence of facilitator influence, albeit unconscious."[6]

Gina Green, Director of Research for the New England Center for Autism, reviewed 15 independent evaluations involving 136 individuals with autism or mental retardation who had been taught to communicate by FC. In no case did an investigator confirm FC.[7]

The American Psychological Association takes the following position:

> [I]t has not been scientifically demonstrated that the therapists are aware of their controlling influence…. Consequently, specific activities contribute immediate threats to the individual civil and human rights of the person with autism or severe mental retardation…. THEREFORE, BE IT RESOLVED that APA adopts the position that facilitated communication is a controversial and unproved communicative procedure with no scientifically demonstrated support for its efficacy.[8]

That doesn't mean that the proponents of FC agree. They argue that the impartial investigations are invalid because "quantitative, objective testing would undermine the confidence of the communicator, place undue pressure on him/her, and introduce negativism that would destroy the communicative exchange."[9]

Syracuse University hosts the Facilitated Communication Institute, an institution founded by Douglas Biklen, Professor of Special Education.

It is astounding to believe that a respectable academic institution would choose to ignore four hundred years of scientific protocol and continue to profess unsubstantiated hypotheses. Professed hypotheses should be tested according to scientific principles. When they fail the tests, they should be abandoned in favor of new hypotheses.

The University of Vermont is another academic institution that chooses to ignore basic scientific protocol. It supports an organization called the Vermont Facilitated Communication Network.

> The mission of the Vermont Facilitated Communication Network is to support the use of facilitated communication (FC) in Vermont by providing education, training and technical assistance, developing resources, disseminating information, and guiding the development and use of best practices.[10]

When there is hesitation in abandoning hypotheses that fail basic science testing, look for a motive. Training and tools that go along with FC might be sold to believers at a significant profit. Meanwhile, people such as Anna's father are being accused of crimes they never committed. The potential for harm could be far greater than the potential benefit.

One might wonder why the experimenter should not expect Anna's typed answer to always be what she sees. If the triple (x,y,z) is (horse, lampshade, cow) or (horse, lampshade, lampshade), why isn't that enough to persuade us that FC is a hoax? Strangely enough, the answer comes from thinking about the *Galton Board*.[11] Sir Francis Galton, the nineteenth-century English geneticist, constructed a board filled with pegs arranged quincunically, like the dots on a die's five-face, with a funnel at the top and chambers at the bottom, as in the illustration on the next page. I never understood the full intent of Galton's demonstration before thinking about the facilitative communication experiment, although I had seen illustrations and demonstrations of the Galton Board many times before.[12]

Drop balls into the funnel at the top—say, drop a hundred balls—and watch how they fall and distribute themselves into the chambers at the bottom. You might guess that they will stack in the chambers in such a way as to define a bell-shape curve similar to the one on page 223 in Chapter 14. Perform this experiment a hundred times, and you will be convinced that the next time you perform it, you will see a bell-shape curve form at the bottom.

But why?

Galton's board had a purpose that seems to have been lost in many books covering it. The Galton Board constructed by the Science Materials Center comes closest to the right idea, but its representation in Mark Kac's 1964 *Scientific American* article titled "Probability" is misleading.[13] Galton's point is to demonstrate that physical events ride on the tailwinds of chance. The ball falls to the first pin and must decide whether to fall left or right. Being unstable in its decision, it flips an unbiased coin; heads go left, tails go right. Whichever way it goes, it falls to the next pin and must decide all over again. What does this really mean? Imagine the perfect Galton Board as one in which the balls always fall directly on the absolute tops of pegs. What makes the ball fall to the right or left? We said that it was a coin flip. But that flip could be determined by a butterfly flapping its wings over the Pacific or a cow farting in an Idaho cornfield.

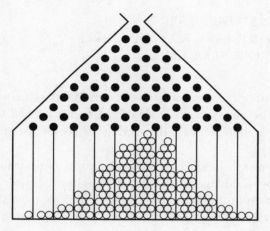

A bounce off the top peg goes left because … well, we don't exactly know why—just because something remotely connected happened, an undetectable air current or vibration. But it represents one of the many unaccounted measurements that imperceptibly affect the outcome. It rolls off the peg to fall directly over the next peg down and, this time, goes right because … who knows?

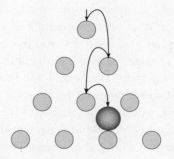

The 50-50 chance of going left or right causes the build-up of the bell-shape curve. Counting the number of ways the balls can fall proves this. Suppose that a ball is dropped and we mark its descent by the letters L and R to indicate bouncing to the left or right. We would then have the following possible outcomes:

LLLL
LLLR, LLRL, LRLL, LRRR
LLRR, LRLR, LRRL, RLLR, LRLR, RRLL
LRRR, RLRR, RRLR, RRRL
RRRR

There are more combinations of mixed letters than nonmixed letters, and because there is an equal chance for the ball to go left or right, there is a tendency for it to end in the center slot of the Galton Board.

Every event in nature has to account for a vast number of indeterminate possibilities. The toss of a die might strongly depend on its initial position in the hand that throws it and more weakly depend on sound waves of a voice in the room. Just two external modifiers guide the die to its resting position. How it strikes the table, the precision of its balance, how it rolls off the hand, and the elasticity of its collision with the table influence which side faces up when it comes to rest.

Return to the experiment designed to accept or reject FC. There is a tendency to argue that Anna should be typing exactly what she sees. Why not? After all, if FC truly works, why would Anna type *cow* when she was seeing a lampshade? One answer is that Anna is human. Humans don't always focus on their tasks. Anna might have seen a lampshade at the very moment her mind wandered to thoughts of visiting her grandparents who live on a farm. How did her mind wander? Hunger, perhaps. Perhaps the tester looked like her grandmother, or perhaps the lampshade itself reminded her of one that sits by the bed of her grandmother. Or perhaps FC doesn't work all the time, but works often enough to be of value. Perhaps it is so sensitive that that proverbial butterfly over the Pacific interfered with FC's effectiveness at a critical moment. That intrusive butterfly makes prediction difficult. Decisions to accept or reject hypotheses are not isolated from all external events; if they were, all results would be determined by a finite number of conditions together with the initial state of the event; we would then know the decision without having to perform an experiment.

CHAPTER 16

Dr. Mortimer, I Presume?

Plausible Reasoning in Science and Math

[W]hile it is often convenient to speak of propositions as certain or probable, this expresses strictly a relationship in which they stand to a corpus of knowledge, actual or hypothetical, and not a characteristic of the propositions in themselves.

—*John Maynard Keynes,* A Treatise on Probability

The Hound of the Baskervilles opens with Sherlock Holmes seated at a breakfast table with his back toward Dr. Watson, who is examining a walking stick that an unidentified visitor had left behind the night before when both Holmes and Watson were out. The stick had a broad silver band with the inscription "To James Mortimer, M. R. C. S., from his friends of the C. C. H., 1884."

Holmes asks Watson what he made of it. The dialogue between Holmes and Watson begins.

"How did you know what I was doing? I believe you have eyes in back of your head."

"I have, at least, a well-polished silver-plated coffee-pot in front of me," said he. "But tell me, Watson, what do you make of our visitor's stick? Since we have been so unfortunate as to miss him and have no notion of his errand, this accidental souvenir becomes of importance. Let me hear you reconstruct the man by an examination of it."

Watson deduces that Dr. Mortimer is a successful, well-esteemed, elderly country doctor, "who does a great deal of his visiting on foot." Holmes harnesses his characteristic conceit for a moment and concedes to Watson, "It may be that you are not yourself luminous, but you are a conductor of light. Some people without possessing genius have a remarkable power of stimulating it, I confess, my dear fellow, that I am very much in your debt."

But then Holmes proceeds to take the stick and examine it himself.

"Interesting, though elementary," says he. "There are certainly one or two indications upon the stick. It gives us the basis for several deductions." He goes on to say, "I'm afraid, my dear Watson, that most of your conclusions were erroneous. When I said that you stimulated me I meant, to be frank, that in noting your fallacies I was occasionally guided towards the truth. Not that you are entirely wrong in this instance. The man is certainly a country practitioner. And he walks a good deal."

Holmes takes the "C. C. H" in the inscription to mean Charing Cross Hospital. Why? He simply says, "The words 'Charing Cross' very naturally suggest themselves." But it gives him a working hypothesis, a place to start. And from this he concludes that his visitor is "a young fellow, amiable, un-ambitious, absent-minded, and the possessor of a favorite dog." He further describes the dog "roughly as being larger than a terrier and smaller than a mastiff."

Of course, the amazing and conceited Holmes is correct on all accounts. It turns out that Dr. Mortimer has a curly-haired spaniel. Holmes's brilliant deductive capabilities are almost machine-like, once they get going; his arrogance supports the machine that carries his self-assured deductions from one to the next. Most persons form opinions far more slowly and with far less certainty than Holmes did. But Holmes's deductive logic is an example of how to fit facts together to form a likely truth. It shows how to fit bits and pieces of facts together into a deductive argument to form a likely conclusion, a conclusion that is no further from the absolute truth than the facts are.

A district attorney's understanding of what it means to deduce guilt—say, in a murder case—leaves me with chilling skepticism. Gary Graham was found guilty of murder by a Texas jury and was sentenced to death on June 21, 2000. The sole evidence of his guilt was an eyewitness account of a woman who identified him in a police line-up. She claimed to have seen Graham shoot a man in a parking lot. She was forty feet away, in her car; it was at night. She even claimed to have seen Graham pull the trigger. Two other eye-witnesses, who claimed that Graham was not the man they saw, were not permitted to testify because they had changed their testimony. Aside from the fact that police line-ups have problems (for example, he was the only member of the line-up with a mustache), how could anyone be so sure about a face that was seen nineteen years earlier? The eyewitness was adamant about the fact that after nineteen years she was still sure.

Penelope would never have recognized her own husband Odysseus on his return from the Trojan War, had it not been for a series of clues and help from her old housekeeper and his nurse, Euryclea, who recognized him by a scar on his thigh. He was a stranger to Penelope, and, though charmed by him, she did not recognize him. Euryclea is told to wash the wanderer's feet, a role of good hospitality to wandering strangers of good breeding. She recognizes the scar but uses other information to make her deduction. The stranger is of the same age and looks like her master. While bathing the stranger's feet, she laments the absence of her master to assess the stranger's reaction.

The word *deduction* means either the act of deducting or the act of deducing. The first meaning suggests the act of "taking away," while the second suggests the act of "leading from." But they take parallel courses when we think that elimination of possibilities leads to a conclusion. Again, take that greatest of all detectives. Given mental stimulation, he can get along without artificial stimulants, but otherwise, only morphine or cocaine will do. In *The Sign of Four*, Sherlock Holmes amuses himself by showing off his deductive powers to Watson; he tells Watson, "Observation shows me that you have been to the Wigmore Street Post Office this morning, but deduction lets me know that when you were there you dispatched a telegram."

How did Holmes know this? Simple deduction. He observed reddish dirt in the instep of Watson's shoe. He knew that the pavement in front of the post office was being redone and temporarily had been filled with reddish dirt that did not exist anywhere else in the neighborhood. "So much for observation. The rest is deduction," he said.

How did he deduce the telegram? He knew that Watson did not write a letter that morning, as he sat opposite him all morning. He also knew that there were a thick bundle of postcards and stamps on his desk. What else could Watson have gone to the post office for but to send a wire?

In triumphant conclusion, he says, "Eliminate all other factors, and the one which remains must be the truth."

Is this deducing or deducting? Perhaps they are one and the same. But the distinction between observation and deduction leads one to believe that there might be two paths to knowing; one might be the path of reason or internal vision, and the other might come from observation and the senses. This division is clearly seen in the differences between mathematical and scientific knowledge.

Holmes seems to interpret deduction as elimination with a sense that truth lies somewhere in the pile of possibilities. He assumed that Watson had a reason to go to the post office and that there are only three reasons: to buy stamps, to mail a letter, and to dispatch a telegram. But surely there are many other reasons to go to a post office. Perhaps he was just walking along Wigmore Street? Possibly he was secretly meeting someone nearby?

Mathematics would never tolerate Holmes's type of deduction. In David Aubern's play *Proof*, Hal, a professor of mathematics studying a proof of a theorem, says that he cannot find anything wrong with the proof, so it checks out.

> HAL: I've been over it twice with two different sets of guys, old geeks *and* young geeks. It is weird. I don't know where the techniques come from. Some of the moves are very hard to follow. But we can't find anything wrong with it! There

might be something wrong with it, but we can't find it. I have not slept. It works. I thought you might want to know.

Hal is essentially saying that if he cannot find anything wrong, it must be true. This is logically equivalent to saying that if it is not true, then he can find something wrong. But wait. No wonder Alice has trouble arguing in Wonderland. Her thinking depends on factual truth, semantics, and conventional logic.

> "Then you should say what you mean," the March Hare went on.
> "I do," Alice hastily replied; "at least I mean what I say, that's the same thing, you know."
> "Not the same thing a bit!" said the Hatter. "Why, you might just as well say that 'I see what I eat' is the same thing as 'I eat what I see!'"

Alice does not distinguish between logical inference and semantics, so how can she distinguish between what she says and what she means?

Isn't the Hatter right? Though *Alice's Adventures in Wonderland* is an exhibition of Lewis Carroll's satire and verbal wit, it is also a repository for examples of factually preposterous statements and wacky logic. How can we find fault with the Cheshire cat's logic, when he lives in a world where he can slowly vanish, "beginning with the end of his tail and ending with his grin"? The cat asserts that dogs are not mad and that he is not a dog, to conclude that he is mad. The problem is that the madness of nondogs does not follow from non-madness of dogs, unless, like him, you are in Wonderland.

It's hard to imagine a world where logic can be twisted to admit that "x implies not y" implies "not x implies y." In the world of Tweedledee and Tweedledum, it is easy to imagine logical connections between nonsensical subjects.

> "I know what you're thinking about," said Tweedledum: "but it isn't so, nohow."
> "Contrariwise," continued Tweedledee, "if it was so, it might be; and if it were so, it would be; but as it isn't, it ain't. That's logic."

We develop powers of deduction through our experiences with cause and effect: Something happens, so something follows. If Watson had been at the Wigmore Street Post Office, he would have reddish dirt on the instep of his shoe. But reddish dirt on the instep of his shoe does not imply that he was at the post office, unless it was the only place where he could have picked up the dirt. Holmes is careful to mention that the reddish dirt does "not exist anywhere else in the neighborhood." The premise Holmes goes on is that Watson went to the post office, and if he did not go for stamps or to mail a letter, then he must have gone to send a telegram.

Though a statement such as "All reddish dirt in the neighborhood is in front of the Wigmore Street Post Office" might not be true, Holmes attached some measure of truth to it. To be sure that such a statement is true, Holmes would have had to check all the reddish dirt in the neighborhood. What he meant was that most reddish dirt in the neighborhood is in front of the Wigmore Street Post Office. It's like being persuaded that all swans are black because most things that are not black are not swans.

We often use our feelings of truth in everyday life when our arguments contain statements that are probably true. For example, take Holmes's conclusion that Watson went to the Wigmore Street Post Office to send a telegram. Holmes thinks he made a deduction. But did he really? It was based not on deductive logic, but rather on the following weak probable inference:

Watson did not go to buy stamps.
Watson did not write a letter that day.
Therefore, Watson went to send a telegram.

A leap is made: In the late nineteenth century, the Wigmore Street Post Office sold stamps, received letters, and sent telegrams. A person would have had no other business going to the Wigmore Street Post Office in those days.

This kind of argument is different than Aristotle's famous syllogism:

All men are mortal.
Socrates is a man.
Therefore, Socrates is mortal.

Holmes assumed that Watson would go to the Wigmore Street Post Office to do one of only three things. That might be likely but is not absolutely certain. We need a measure of likelihood and all measures are numbers. So what number will persuade us that Watson did send a telegram that day?

Bacon, who was a better philosopher than scientist or Lord Chancellor, suggested that the natural sciences had to give up the authority of writings from antiquity, including Aristotle's, to become more respectable. He fervently opposed the classical notion of understanding nature through strictly deductive logical inference. Though his idea of induction seems naïve from a modern standpoint, he is given credit for the method of inductive enquiry used by today's scientists, partly because he was such a severe critic of the deductive form of science. His aphorisms show the flavor of his attacks.[1]

> The subtlety of nature is greater many times over than the subtlety of the senses and understanding; so that all those specious meditations, speculations, and glosses in which men indulge are quite from the purpose, only there is no one to observe it.

> The syllogism is not applied to the first principles of sciences, and is applied in vain to intermediate axioms, being no match for the subtlety of nature. It commands assent therefore to the proposition, but does not take hold of the thing.

> The syllogism consists of propositions, propositions consist of words, words are symbols of notions. Therefore if the notions themselves (which is the root of the matter) are confused and overhastily abstracted from the facts, there can be no firmness in the superstructure. Our only hope therefore lies in a true induction.

By contrast, Tolstoy warned against imprudent generalization, suggesting that we humans see small phenomena as the causes of greater phenomena:

> "Whenever there have been wars, there have been great military leaders; whenever there have been revolutions in states, there have been great men," says history. "Whenever there have been great military leaders there have, indeed, been wars," replies the human reason; but that does not prove that the generals were the cause of the wars, and that the factors leading to warfare can be found in the personal activity of one man.
>
> The peasants say that in the late spring a cold wind blows because the oak-buds are opening, and, as a fact, a cold wind does blow every spring when the oak is coming out. But though the cause of a cold wind's blowing just when the oaks are coming out is unknown to me, I cannot agree with the peasants that the cause of the cold wind is the opening of the oak-buds, because the force of the wind is altogether outside the influence of the buds.[2]

If the beating of a butterfly's wings over the Atlantic can cause a hurricane in the Pacific, we should think that almost nothing could be measured with enough precision to predict future events without error. A hydrogen atom moves from one energy state to another by either absorbing or emitting photons, but this behavior is random; whether it emits or absorbs is as much a matter of chance as flipping a coin. If states of hydrogen are determined by the whims of chance, what chance does the rest of the world have in being predictable? On the one hand, we tend to be satisfied with plausible assurance that an event will happen or credible faith that something is true. The 95 percent certain rule might be acceptable in scientific studies when no direct deductive arguments are possible, but how acceptable would it be if exact deductive logic could be possible?

Take the case of the unsolved mathematical problem originally stated in a letter from the Prussian mathematician Christian

Goldbach to Swiss mathematician Leonhard Euler in 1742 that every even integer greater than 2 can be written as the sum of two primes.

$$4 = 2 + 2$$
$$6 = 3 + 3$$
$$8 = 3 + 5$$
$$10 = 5 + 5 = 3 + 7$$
$$12 = 5 + 7$$
$$14 = 3 + 11 = 7 + 7$$
$$16 = 3 + 13 = 5 + 11$$
$$18 = 5 + 13 = 7 + 11$$
$$20 = 7 + 13$$

There are an infinite number of even numbers, and we would like to know whether every one of them could be written as a sum of two primes. In 1742, a finite analog of the problem would have been interesting. If Goldbach's letter had asked if every even number less than a billion is the sum of two primes, the problem would have been interesting, but not as interesting as the same question without the size limit. It would not be too difficult to have a computer make a list of even numbers under a billion and to work out their decompositions into sums of primes, but what about ten billion or a billion billion? The sum of any two prime numbers greater than 2 is even, but can all even numbers less than a billion billon be written as a sum of two primes? We could try to write a program that checks: Given an even number $2n$, the program would run through pairs of primes less than $2n$ checking that their sum equals $2n$. The problem is that we don't know all the primes; at the moment, we don't know enough of them.

The answers to many questions in science are known only to 95 percent certainty. So why not be happy with 95 percent certainty for the Goldbach conjecture? It would be silly to ask if the Goldbach conjecture is true for 95 percent of all even numbers because there are infinitely many even numbers, and 95 percent of infinity is still infinity. We might take a random sample of one thousand even numbers as high as a billion billion and ask how many can be written as a sum of two primes. But unlike experiments with go chips, we do not know the probability of picking an even number equaling the sum of two primes. We hope that that probability is one. Is there merit in sampling thousands of even numbers less than a googolplex (the number $10^{(10^{100})}$) to test the conjecture at the 95 percent level? The largest known prime is $2^{20996011} - 1$, a 6,320,430-digit number that would fill a good chunk of the Manhattan telephone book.[3] It is less than $10^{(10^7)}$, far less than googolplex.

Can there be any merit in using sampling methods to prove statements of pure mathematics? Sure. Instead of checking even numbers on an ordered list such as 4, 6, 8, 10, 12, ..., we can save time by randomly looking at even numbers and checking whether they can be written as sums of two primes. Take the question of Mersenne numbers, named in honor of seventeenth-century mathematician Marin Mersenne. They are numbers M_n defined by $M_n = 2^n - 1$, where n is an integer greater than one. If you examine these numbers in increasing order of n, you find that some of them are prime and others are not. Are there infinitely many Mersenne primes? We don't know. We also don't know how the Mersenne primes distribute themselves between the Mersenne numbers. Take a look at how the primes distribute themselves in the list of the first seventeen Mersenne numbers:

$M_2 = 3$—prime
$M_3 = 7$—prime
$M_4 = 15 = 3 \times 5$
$M_5 = 31$—prime
$M_6 = 63 = 3 \times 21$
$M_7 = 127$—prime
$M_8 = 255 = 5 \times 51$
$M_9 = 511 = 7 \times 73$
$M_{10} = 1023 = 11 \times 93$
$M_{11} = 2047 = 23 \times 89$
$M_{12} = 4095 = 5 \times 819$
$M_{13} = 8191$—prime
$M_{14} = 16,383 = 3 \times 5,461$

$M_{15} = 32,767 = 7 \times 4,681$

$M_{16} = 65,535 = 5 \times 13,107$

$M_{17} = 131,071$—prime

One might think, as Fermat did, that numbers of the form $2^{2^n} + 1$ are prime; after all, the first four such numbers are prime.

$$2^{2^1} + 1 = 5, \quad 2^{2^2} + 1 = 17, \quad 2^{2^3} + 1 = 257, \quad 2^{2^4} + 1 = 65,537.$$

But in 1732, Leonard Euler showed this:

$$2^{2^5} + 1 = 641 \times 6,700,417.$$

The moral here is that one should be very suspicious of inductive arguments in mathematics. On the other hand, they certainly have their place and can be very useful when forming plausible impressions of truth.

Early Greek mathematicians, as far back as the Pythagoreans, were interested in those numbers equal to the sum of all positive, proper divisors.[4] They called such numbers *perfect numbers*, possibly because they attached such mystical powers to numbers. Even Saint Augustine attributed the number of days it took God to create the world to the perfection of the number 6.

If you examine the first four perfect numbers—6, 28, 496, and 8,128—you might conjecture that the nth perfect number should have n digits, but you would be wrong: The next perfect number is 33,550,336. You might then conjecture that the last digits alternate between 6 and 8, but, again, you would be wrong: The next perfect number is 8,589,869,056. So, you might alter your conjecture to say that all even perfect numbers end in either a 6 or an 8. And now, you would be right. You might wonder about how these numbers are growing; from one to the next, there is a huge increase in size. The next perfect number is 137,438,691,328. Indeed, they are very rare.

Are there infinitely many perfect numbers? We don't know. But Euclid proved that $2^{n-1}(2^n - 1)$ is a perfect number whenever $2^n - 1$ is a prime number.[5] The Mersenne number $M_{13466917} = 2^{20996011} - 1$ is prime, which would say, by Euclid's proposition, that $2^{20996010}(2^{20996011} - 1)$ is a perfect number. This perfect number (which is only the fortieth) has more than ten thousand billion digits.

Of course, there might be perfect numbers that are not of the form $2^{n-1}(2^n - 1)$, but such a number must be even, so this opens up two new questions. First, are there perfect numbers that are not of the form $2^{n-1}(2^n - 1)$? Second, are there any odd perfect numbers? Euler proved that every even perfect number is of the form $2^{n-1}(2^n - 1)$ with its $(2^n - 1)$ factor a Mersenne prime. That proof, together with Euclid's proposition, tells us that every even perfect number is of the form $2^{n-1}(2^n - 1)$; likewise, every number of the form $2^{n-1}(2^n - 1)$ with $(2^n - 1)$ prime is perfect. The second question is still an open problem: No odd perfect number has ever been discovered, and neither has a theorem contradicting its existence.

We have a fair number of ancient unsolved problems. In addition to the questions of if there are infinitely many perfect numbers, and if there are any odd perfect numbers, there is Goldbach's conjecture, for one: are there infinitely many Mersenne primes? The list could go on, but the central question we might ask is this: How do we form opinions about any answers? The answer is that we resort to the same techniques we use when forming opinions in our everyday lives: We base our judgments on experience.

Experience is mostly circumstantial; opinion is influenced by culture and conditioning. Though Anna's story contains almost no information from which to draw reasonable conclusions, the telling of the story conjures opinion. If you know something about Anna's story, you might have a point of view quite opposed to the biases given here. Between 1990 and 1995, at least fifty similar cases were tried in the United States involving accusations of sexual abuse by adults through messages produced by FC. Take another example: A controversial article by two researchers from the University of Chicago and Yale University connects the drop in the crime rate to abortion for the decade of the 1990s. The reason given is that there are fewer "unwanted" children in the world and that unwanted children usually are born to families that do not give children the nurturing that they need to grow up and be productive members of society. Two economists, Steven Levitt and John Donohue III, conducted the research. They contend that the Supreme Court case *Roe* v. *Wade*, which legalized abortion in 1973, reduced the number of unwanted children and, thus, reduced the number of children who would have been more likely to become criminals. We have here a very different case for persuasion. We are asking about the truth of a relationship between two statistics.

One expects a jury to be unbiased before hearing evidence for or against conviction. Yes, it *should* have no opinion until all the evidence is presented, but real juries are made of humans with good intentions on the one hand and human faults and limitations on the other. No jury will be able to keep opinions from forming. Hit the jury with factual evidence in favor of the prosecution and it favors guilt. Hit it with evidence in favor of the defense, and a verdict of not guilty will be favored. Like a pinball falling through the maze of pins, it scores one way or the other.

There are those who refuse to be persuaded, even in the face of enormous evidence. Conspiracy theorists interpret evidence to support their view. Some people believe that the NASA moon landing was a hoax. In fact, conspiracy theorists were able to convince the Fox Broadcasting Company to air a special broadcast on the subject. And although the broadcast disclaimed any bias, it withheld information that would have been useful to make reasonable judgments against the conspiracy theory. The program claimed that 20 percent of all Americans believe we never landed on the moon. When a program on a major television station puts forward a convincing view in support of a conspiracy, viewers tilt toward supporting the conspiracy.

The theory contends that the entire mission was filmed in a studio, that Neil Armstrong never went to the moon, and that thousands of NASA scientists have been part of the conspiracy for the past thirty-two years. And what is the evidence? Photographs of Neil Armstrong show two shadows, even though the sun should have been the only source, so the other source must have been studio spotlights. The photographs show no stars, no blast-off crater, and a waving American flag.

Surely, if the mission were a hoax, at least one of the thousands of NASA scientists would have thought of these faults. It turns out that every one of the objections has a scientific explanation that escaped the conspiracy theorists: There should have been more than one source of light (the reflective surfaces of Earth and the moon), there should not have been stars visible in the photograph (stars need a time exposure to appear in a photograph), the Lunar Landing Module should not have left a visible crater (moon dust is only a couple of inches thick) and flags wave when their poles are disturbed. In fact, a hoax would have been more likely, had there been only one light source, stars in the photograph, a blast-off crater, or an unmoving flag. Perhaps it is easier to believe the mission to be a hoax than to comprehend how difficult it really was. How would Francesco Sizzi, the Italian astronomer who mocked Galileo's observations of the moons of Jupiter, have reacted to the NASA pictures?

We have seen that what happens by chance might also tend to mathematical order and pattern. Recall from Chapter 15 the balls randomly dropping down the Galton Board, bouncing from peg to peg left and right with equal likelihood, yet tending to accumulate in a pattern. One could design boards so balls bounce with unequal likelihood from peg to peg and predict the pattern of their accumulation. Chevalier de Méré and Blaise Pascal could not have imagined matter as huge numbers of atoms and molecules crashing into each other at great speeds, bouncing in confusing directions with no apparent pattern. They could not have imagined heredity as the transmission of complex strings of messages written with just four letters and evolution due to random errors in copying those messages. They were thinking of a pair of dice and the smallest number of times a person must throw them to have a better than even chance of getting a double six. It would seem that the throwing of dice should have nothing to do with predicting weather that might be caused by a butterfly's wings over the Atlantic, and yet, if we could see the infinitesimal connections of accumulated changes caused by those wings, we might look to the butterfly to predict the weather.

A pair of dice lands on boxcars. Why? Not because there is a one-in-thirty-six chance and it was time for boxcars to show. The one-in-thirty-six chance is just a mathematical reason, a brilliant macromodel designed to predict an event influenced by that infamous butterfly. We are so accustomed to thinking in terms of classical mechanics, of cause and effect, that we believe that there should be an undetectable moment when the dice are held in one of a huge number of possible positions in the palm of a hand; or while spinning in two directions through the air, continuously affected by countless currents; or striking a surface and bouncing again and again, that inevitably determines boxcars. The logic we use to predict positions of the planets or where a rocket will fall does not apply to the mechanics of small things. We can safely say where the dice will fall if the toss is not too wild, but not how. But we will never know the moment the dice decide to show boxcars. We can say that if the dice are thrown thirty-six times, there is a chance that they will show boxcars once. Like all good questions, the one Chevalier de Méré asked in the winter of 1654 didn't end simply with an answer. We now know that it takes twenty-five throws to have favorable odds of throwing boxcars, but a whole new way of thinking about prediction in our enormously complex world of indeterminate events came to us by way of de Méré's question.

Conclusion

Sometimes when, for a time, we think hard about some subject and then "dismiss it from our minds," this subject seems to reappear later, in an improved and expanded form, just as though it had descended to a sub-conscious level of the mind, and had there profited by association with an unrecognized mixture of other ideas in a sort of unconscious reverie.

—Warren Weaver, Lady Luck

Truth comes in many forms. Having set out the process of per-suasion in three contrasting forms of logic and classical proof; the math behind the surprises of infinity; and the plausible reasoning, we are still left with the mystery of human creativity. All this logic and math and science is, after all, something only we do. How do we come up with scientific hypotheses in the first place? Why do we care about the way some things go on and on and on? Why should we play around with triangles trying to prove things about them? There has been a debate on the nature of reason for centuries. In the twentieth century (led by thinkers such as Bertrard Russell, Ludwig Wittgenstein, and Kurt Gödel) we discovered that logic is rooted in language as well as mathematics. And language always involves human sense and intuition because it cannot exist without the rainforest of human culture. After two and a half millen-nia of Western thought we might think Euclid was robotic in his pursuit of the truth about geometry. But humans are always more than sophisticated computer circuits. Euclid too. He was an explorer inventing mental tools to deal with his environment; and as the gen-erations have passed, our experience of the complexity of our world has deepened. So in turn our tools have become complex—in ways Euclid could not have imagined. The pursuit of ultimate truth will go on, in the human mind, and perhaps in artificial intelligence of our invention. Yet anyone who loves to bring nature and humanity closer together will remain a logician of sorts; an observer who can deduce answers; a mathematician who can handle infinity; a scientist who can accept chance. A Euclid in the rainforest.

Nearly every six minutes, a commuter train passes through Harlem on elevated tracks heading north from Grand Central Station. If you were on one of those trains in the last century, you might have seen acres of neglected half-abandoned tenement houses, city-forsaken playgrounds, and streets littered with overstuffed, overflowing garbage bags propping up the local inebriated down-and-out. From the quiet inside of your commuting train, you might have seen cracked asphalt basketball courts enclosed by broken chain-link fences, discarded mattresses, disposed tires, and wide splashes of graffiti covering brick walls of public housing.

On a recent sunny spring afternoon, I was on one of those trains heading north and couldn't help noticing how Harlem was changing, becoming almost as attractive as it must have once been, back in the late nineteenth century. Basketball courts now have maroon rubberized surfaces and new fences. Streets are clean now and are lined with newly planted chestnut trees and ornate replicas of nineteenth-century lampposts. Abandoned tenement houses were demolished, giving a clear view—from the train—of rows of fine nineteenth-century brownstone houses. No graffiti anywhere.

A brawny African American dressed in a business suit with a laptop on his lap sat next to me. As the train slowed down, he pointed out the window.

"Look at Harlem now," he said in a friendly voice, tilting his head downward to see above the reading glasses low on his nose. "One could almost imagine living here."

"I'm amazed at how much it has changed since the last time I passed by," I volunteered. "That was only six months ago, but I didn't bother to look out the window to notice the change."

"Well, I used to live here, believe it or not," he said quietly. It was almost a whisper.

"Really? When was that?" I asked, puzzled that he would confess this personal piece of information to a total stranger.

"Long time ago," he answered. "You see that building being raised?" he asked, pointing west. "That's where I lived."

He pointed to the steel skeleton of an office building under construction, surrounded by lifts and cranes. From that view, the future of Harlem looked promising. After a long pause, in which neither of us said anything, he continued, "The new Harlem economy." I detected a tinge of criticism in his voice.

"Where will all those people go when the rents start to go through the roof?" he rhetorically asked, articulating each word as if he were dictating to someone taking notes. He shook his head and looked down to continue work on his laptop. I began to read the book I had brought along for the trip back to New Haven.

As the train picked up speed and passed over a bridge into the Bronx, I fell into the distraction of peeking at my neighbor's screen, which was filled with multicolored graphs and charts suggestive of ambitious mathematics. My head imperceptibly moved so my snooping eyes could turn as far right as possible to invade the private text and read his screen. My movements were not as delicate as I thought, for at the next moment, my neighbor starkly turned and said, "Are you interested in what I am doing?"

"I'm sorry. I—I was just curious," I murmured. "I couldn't avoid noticing the mathematics on the screen and only wondered because I'm a mathematician."

"Oh," he said, regaining his friendly tone. "I'm a statistician. I'm a consultant for gambling commissions, state lotteries, and racetracks—anything that involves gambling odds."

"So, what are you working on?" I asked. For the next half-hour, my neighbor rapidly talked about his graphs and equations, pointing to them, alternately looking at the screen and me, his head lifted to see through his reading glasses or tucked to his chin to see over the top. His words were familiar, but their meaning seemed perplexingly absurd, like jigsaw pieces of a dream that seem true until one wakes. I understood very little but asked enough questions to barely feel

I understood something when his cell phone rang to stop his lecture. He talked on the phone while packing his laptop into a briefcase monogrammed U.B., getting ready to get off at Stamford, the next stop.

The train came to a stop and the doors opened. With a wave of his hand, he bid a farewell.

My mind raced backward when the doors closed, overflowing with daydreams of friends and acquaintances I hadn't seen or heard from in forty years. As the train made its way north to New Haven, stopping at almost every local stop along the way, all I could think of were old encounters with Roger and Jesus through rainforests of recollections and stories I had long forgotten, particularly dazed by one of those reveries that paint new colors on old memories.

I could almost smell the coffee, lemon, and thyme as the trance took over. I could hear the two-string mandolin and the coffee woman humming her song in the Cabruta marketplace. There I was, buying a mango and sunglasses, just as I had done forty years ago, when suddenly this reverie jumped to a brilliant memory of one forgotten event just before crossing the Orinoco. Just south of the ferry dock, alongside giant swaying mangroves and arboreal ferns, amid the watching white faces of opossums hanging from trees, stood a grass hut. A thin black woman dressed in a long dress made from brightly colored patches of cotton, with a tiger tooth dangling on a leather cord around her neck, sat on a palm log, humming a melody and shaking her head as if in a trance. Her hair was pinned up and head-dressed with the same brightly colored patchwork of cotton under an agal. I learned from my company that she was a mambo, a priest of a powerful voodoo god, Damballah, the god of all gods.

"Don't worry," Renaldo assured us, "It's just Delu. She's here to tell the fortunes of people about to cross the river."

"Damballah, all good, never do bad deeds," said Delu as we approached. Her humming stopped just long enough to say, "Voodoo worshipers could ask favors from other gods, like Christians ask favors from saints, but they must first come to Damballah for okay and power."

"But don't voodoo worshipers believe in evil spirits that can be summoned to do evil work through their powers of charms and sorcery?" Roger asked. In Spanish, his s-lisp was even more pronounced. But in my reverie, his speech impediment had disappeared.

"Surely, none of us believes…," Jesus replied.

"Voodoo magic may be abused," Delu cut in rhythmically. "But this mambo never send for evil spirits."

"Yeah," said Roger, "but I heard that a priest could chant the right prayer while tearing the tongue out of a red rooster, smear the blood on a wall and add the feathers from the throat, stick a needle in a rag doll somewhere in Venezuela, and your ex-girlfriend's boyfriend in London will grow a wart at the end of his nose."

Delu laughed, showing few but large yellow and brown teeth.

Roger cracked his knuckles and gave Delu three bolivars for a private reading of his destiny. Uncharacteristic of him. Waiting outside the hut, I imagined Delu mixing a solution of blood with a feather that could have been torn from the throat of a red rooster, listening to Delu chant and intermittently call out bewilderingly irrelevant numbers quite loudly. When he came out, I asked him about the numbers; he did not answer, but Jesus speculated that it had something to do with the speed of the spinning blood potion. Curious, he and I both imagined that there really was a spinning blood potion. After some time had passed, he said that the shape of the top surface of the blood was "that of a bowl cupped upward."[1]

"But, Roger, what is your destiny?" I asked, while pondering how closely mine was connected with his.

"Oh, the usual nonsense: that I will marry a rich and beautiful woman, have lots of adventurous kids, live forever," he said. Then, after a long pause, he added, like some character from a Rudyard Kipling story, "And fear not the python, but beware the swarming bugs—they bite."

He was abnormally silent as we crossed the river on the ferry-raft with Renaldo and Jesus. All of us wanting to be Euclids in the rainforest.

Proof That All Triangles Are Isosceles[1]

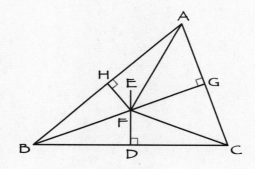

Let *ABC* be any triangle. Bisect *BC* at *D*, and from *D*, draw *DE* at right angles to *BC*. Bisect the angle *BAC*.

1. If the bisector does not meet *DE*, they are parallel. Therefore, the bisector is at right angles to *BC*. Therefore, *AB* = *AC* (that is, *ABC* is isosceles).

2. If the bisector meets *DE*, let them meet at *F*. Join *FB* and *FC*, and, from *F*, draw *FG* and *FH* at right angles to *AC* and *AB*.

The triangles Δ*AFG* and Δ*AFH* are equal because they have the side *AF* in common, and the angles ∠*FAG* and ∠*AGF* are equal to the angles ∠*FAH* and ∠*AHF*. Therefore, *AH* = *AG*, and *FH* = *FG*.

Again, the triangles Δ*BDF* and Δ*CDF* are equal because *BD* = *DC*, *DF* is common, and the angles at *D* are equal. Therefore, *FB* = *FC*.

Again, the triangles $\triangle FHB$ and $\triangle FGC$ are right-angled. Therefore, the square on FB equals the squares on FH and HB, and the square on FC equals the squares on FG and GC. But $FB = FC$, and $FH = FG$. Therefore, the square on HB equals the square on GC. Therefore, $HB = GC$. Also, AH has been proved to be equal to AG. Therefore, $AB = AC$ (that is, $\triangle ABC$ is isosceles). Therefore, the triangle $\triangle ABC$ is always isosceles.

A Method for Unraveling Syllogisms

Several reasons explain why the conclusion of a syllogism is not immediately clear. First, you might not notice what the syllogism is about. Second, there may be too many negative attributes. Third, the statements themselves might not be listed in their most natural order. Let's consider these reasons, one at a time. (The following is based on the syllogism given on page 53.)

1. First, notice that the entire syllogism is about something. Did you notice that every part of the syllogism is about kittens? That gives us a place to start. Every statement is about connections between attributes of kittens. For example, the first statement means "No kitten that loves fish is an unteachable kitten."

2. Notice that you can translate most statements into positive statements. Statements of the form "No *x* is *y*" translate to "All *x*s are not *y*s." The first translates to "Kittens that love fish are teachable." The statements then become

   Kittens that love fish are teachable.
   Kittens that play with gorillas have tails.
   Kittens with whiskers love fish.
   Teachable kittens do not have green eyes.
   Kittens with tails have whiskers.

   The fourth statement still contains the words *do not*, but you will see that that is okay.

3. Rearrange the statements in the following way:
   Kittens that play with gorillas have tails.
   Kittens with tails have whiskers.
   Kittens with whiskers love fish.
   Kittens that love fish are teachable.
   Teachable kittens do not have green eyes.

From this, you can safely conclude, "Kittens that play with gorillas do not have green eyes" or, equivalently, "No green-eyed kitten will play with a gorilla."

This might not be the way you approached the problem. It is a mathematical problem, and, as with all mathematical problems, there are multiple approaches to a solution. Let's break down the problem, statement by statement. Because we are always talking about kittens, let's eliminate the word *kitten* and abbreviate the attributes as follows:

No kitten that loves fish is unteachable = LOVES FISH implies TEACHABLE

No kitten without a tail will play with a gorilla. = PLAY WITH GORILLA implies TAIL

Kittens with whiskers always love fish = WHISKERS implies LOVES FISH

No teachable kitten has green eyes. = TEACHABLE implies NO GREEN EYES

Kittens that have no whiskers have no tails. = TAILS implies WHISKERS

Hence,

LOVES FISH implies TEACHABLE
PLAY WITH GORILLA implies TAIL
WHISKERS implies LOVES FISH
TEACHABLE implies NO GREEN EYES
TAILS implies WHISKERS

Because every statement is about kittens, we can think of the attributes as simply names for the classes of kittens with such attributes. Perhaps we should make the convention that the name of the class is in brackets ({}), so {TEACHABLE} is the name of the class of teachable kittens, {LOVE FISH} is the name of the class of kittens that love fish, and so on. We can say this:

1. {LOVES FISH} is contained in {TEACHABLE}.
2. {PLAY WITH GORILLA} is contained in {TAIL}.
3. {WHISKERS} is contained in {LOVES FISH}.
4. {TEACHABLE} is contained in {NO GREEN EYES}.
5. {TAIL} is contained in {WHISKERS}.

If we diagram these classes, we see the following.

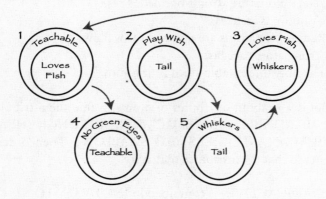

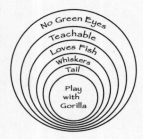

You can now nest these classes, one in another, as indicated by the illustration on the left.

Not all syllogisms are this easy to unravel. Let's take another. Keep in mind that the conclusion of a syllogism might be valid, even though each statement is not factually true.

Atoms that are not radioactive are always unexcitable.
Heavy atoms have strong bonds.
Uranium is tasteless.
No radioactive atom is easy to swallow.
No atom that is not strong is tasteless.
All atoms are excitable, except uranium.

What happens if you try to write statements with the least number of negative components and then organize the statements? You should get something like this:

1. Excitable atoms are radioactive.
2. Heavy atoms are strong-bonded.
3. Uranium is tasteless.
4. Radioactive atoms are not easy to swallow.
5. Strong atoms are tasty.
6. The only unexcitable atom is uranium.

The syllogism is about attributes of atoms. Name the attributes as follows: {EXCITABLE}, {RADIOACTIVE}, {STRONG}, {HEAVY}, {URANIUM}, {TASTY}, {EASY TO SWALLOW}. Then order the statements so that the attributes match up.

4. {RADIOACTIVE} is contained in {NOT EASY TO SWALLOW}.
1. {EXCITABLE} is contained in {RADIOACTIVE}.

6. {NOT URANIUM} is contained in {EXCITABLE}.
3. {TASTY} is contained in {NOT URANIUM}.
5. {STRONG} is contained in {TASTY}.
2. {HEAVY} is contained in {STRONG}.

Illustrate the containment of classes by the diagrams that follow.

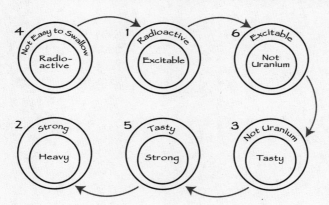

It is now easy to see that we can set these classes into the nest of classes below. And so we see that heavy things are not easy to swallow.

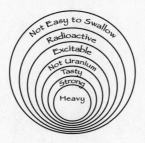

Density of Irrationals on the Number Line

Demonstrations to Show That Between Every Two Rational Numbers There is an Irrational and That Between Every Two Irrationals There is a Rational Number

This might be counterintuitive, but it is easy to show. Assume that the two rational numbers are very close to each other, so the difference is very small. Divide the difference by π and add the result to the smaller of the two numbers. You now have an irrational number between the two original rational numbers.

To show that there is a rational number between any two irrational numbers, a and b, draw lines starting at (0,0) with slopes a and b. These lines are labeled α and β in the illustration below. The grid points in the illustration represent points with addresses (p,q), where both p and q are integers.

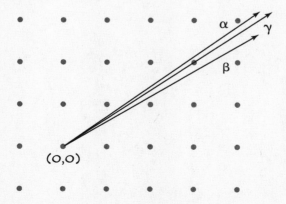

(The slope of a line is its steepness. To calculate it, imagine the line as lying on the hypotenuse of a right triangle whose base is horizontal. Then the slope is the ratio of the height of the triangle divided by its base.)

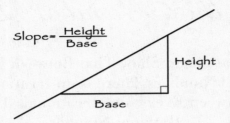

Notice that neither α nor β will ever hit a point (p,q), where both p and q are integers; if either did, it would have a rational slope p/q. Also notice that the angle between α and β fans out and will eventually enclose one of the grid points (p,q), where both p and q are integers. Draw the line γ from $(0,0)$ to the grid point enclosed by the angle between α and β. The slope of γ is p/q, where p and q are integers, and, hence, a rational number between the two original irrational numbers a and b.

Cantor's Demonstration That the Real Numbers Are Uncountable

The argument for why the integers and rational numbers have a cardinality less than that of the real numbers goes as follows:

Every fraction p/q can be placed on a two-dimensional array, addressed by its vertical and horizontal position—q places to the right and p places down—as in the following figure.

$$
\begin{array}{cccccccc}
\dfrac{1}{1} & \dfrac{1}{2} & \dfrac{1}{3} & \dfrac{1}{4} & \dfrac{1}{5} & & \dfrac{1}{q} & \\[2ex]
\dfrac{2}{1} & \dfrac{2}{2} & \dfrac{2}{3} & \dfrac{2}{4} & \dfrac{2}{5} & & \dfrac{2}{q} & \\[2ex]
\dfrac{3}{1} & \dfrac{3}{2} & \dfrac{3}{3} & \dfrac{3}{4} & \dfrac{3}{5} & \cdots & \dfrac{3}{q} & \cdots \\[2ex]
\dfrac{4}{1} & \dfrac{4}{2} & \dfrac{4}{3} & \dfrac{4}{4} & \dfrac{4}{5} & & \dfrac{4}{q} & \\[2ex]
\dfrac{5}{1} & \dfrac{5}{2} & \dfrac{5}{3} & \dfrac{5}{4} & \dfrac{5}{5} & & \dfrac{5}{q} & \\[2ex]
& & \vdots & & & & \vdots & \\[2ex]
\dfrac{p}{1} & \dfrac{p}{2} & \dfrac{p}{3} & \dfrac{p}{4} & \dfrac{p}{5} & & \dfrac{p}{q} & \cdots \\[2ex]
& & \vdots & & & & \vdots &
\end{array}
$$

Start listing these numbers according to their place along the infinite serpentine arrow in the following figure. You now have all rational numbers (with repeats) on an ordered list so that each rational number corresponds to one and only one whole number (if you ignore repeats). You can do the same for negative rational numbers. This is just another way of saying that there are just as many rational numbers as there are integers.

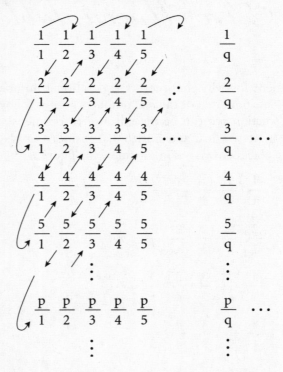

When you try to do the same thing with real numbers, you will find that there is a problem. Recall that real numbers between 0 and 1 are numbers that can be represented by a (possibly infinite) string of decimals. For example, 0.4673904739828983493... is one such number in which the three dots indicate that the digits go on forever. Cantor showed that the set of all real numbers could not be listed. (In other words, the set of real numbers could not be put into a one-to-one correspondence with the integers.) Just try to list real numbers from 0 to 1. One possible list might look like this next figure.

1	0.46739 $\cdots$
2	0.38654 $\cdots$
3	0.03936 $\cdots$
4	0.84534 $\cdots$
5	0.67657 $\cdots$
	$\vdots$

No matter how you arrange these real numbers, there will always be infinitely many that are not on your list. Here is just one: Take the number that you get by reading down the infinite diagonal of the infinite array (diagonal encirclement). The number you get is 0.48937....

n-th digit of *n*-th number on the list

Now construct a new number by changing this diagonal number as follows: If a digit is not 9, add 1 to it. Change any 9 into a 0. In this example, the first digit becomes 5, the second becomes 9, the third becomes 0, and so on, so the newly constructed number is 0.59048.... This number is not on the list. If it were, it would have to be somewhere on the list—say, in the *n*th position, and we would come to the most bizarre situation of having a digit d, the *n*th digit of the *n*th number on the list (the one in the circle) that is both d and $d+1$ at the same time. The only recourse is to admit that the number we constructed (0.59048...) is not on the list. This shows that the set of real numbers is *larger* than the set of integers and, hence, *larger* than the set of rational numbers. The digits along the diagonal go on forever, so infinitely many numbers could have been constructed in this way, giving infinitely many not on the list. (We could have added 2 or 3 or any number between 1 and 9 to the digits of the diagonal number and achieved the same result.) This shows that the cardinality of the set of real numbers is greater than the cardinality of the set of rational numbers.

Chapter 1, *The Search for Knowledge School*

1. Now under attack by Patrick Tierney, who wrote a book titled *Darkness in El Dorado: How Scientists and Journalists Devastated the Amazon* (New York: W. W. Norton & Co., 2002).

2. An isosceles triangle is a triangle with two sides of equal length.

3. This was an easy calculation, based on estimating the slope of the ravine to be about 45°. The force on the cable would have to be at least the weight of the truck times the sine of 45°. $(2 \times \sin(45) = 2 \times .707 = 1.414)$

4. Proclus, *Proclus: A Commentary on the First Book of Euclid's Elements*, trans. (with introduction and notes) Glenn R. Morrow (Princeton, NJ: Princeton University Press, 1992).

5. A *right angle* is an angle measuring 90°. A *right triangle* is a triangle that has one right angle.

6. Mathematical theorems, though unshakably true, may have no unique wording in an informal language such as English. The following is merely one way of wording the Pythagorean theorem: "In any right triangle, the sum of the squares of the lengths of the sides equals the square of the length of the hypotenuse." But what are the meanings of the words in this statement? What is a right triangle, what are its sides, and what is its hypotenuse?

The Pythagorean theorem can be tested over and over again with different objects that fit its conditions. This is not the same as constructing a proof. To be fair, we should put formal mathematical statements into a different category than informal statements about the world. How could we know the real truth about who shot John Kennedy or who President Clinton slept with?

Mathematical statements such as the Pythagorean theorem have definite forms that can be checked. They are conditional; that is, the truth (of the conclusion) depends on whether the objects that they talk about satisfy certain conditions. The theorem states that if $\triangle ABC$ (the object) is any right triangle (the condition), the sum of the squares of the lengths of the sides equals the square of the length of the hypotenuse (the conclusion). The object in question is a triangle; the condition that is to be satisfied is that it has a right angle. If it does, then we may conclude that the sum of the squares of the lengths of the sides of that triangle equals the square of the length of its hypotenuse. It may seem a formidable task to try to untangle what constitutes a proof, yet the task is far easier than you may think.

7. http://www.Guinnessworldrecords.com/content_pages/record.asp?recordid-51998. "Eleftherios Agryopoulos of Greece has discovered five hundred and twenty different proofs of the Pythagorean theorem over a period of eleven years from 1986 to 1997."

Chapter 2, *How to Persuade Jesus*

1. See *Thirteen Books of Euclid's Elements*, trans. T. L. Heath (Mineola, NY: Dover Publications, 1956), 36–37, 185.
2. I recently met someone who visited Cabruta in 1999. This is now an area rich in oil. The town boasts several major hotel chains, a Burger King, and a McDonalds. But in 1960, Cabruta was on the jungle's boundary, with only one dirt road leading in and out.

3. Here is Roger's proof in more detail. Consider any right triangle $\triangle ABC$ for which AB is the side opposite the right angle. Construct a line through the right angle perpendicular to AB, meeting at E. You now have three right triangles: the original $\triangle ABC$, $\triangle AEC$, and $\triangle CEB$. Now, the interesting thing about these triangles is that the area of $\triangle AEC$ plus the area of $\triangle CEB$ equals the area of $\triangle ABC$. This is beyond doubt. It comes from the common notion that the whole is the sum of its parts. So this clearly proves the Pythagorean theorem for these particular shapes, right triangles whose hypotenuses are equal to the respective sides of the original triangle. But how do we know that the theorem is true for other shapes, such as squares? All three triangles are proportional to each other because their corresponding angles are equal. Being proportional also means that ratios of their corresponding sides are also equal. Therefore, $|AC|/|AE| = |AB|/|AC|$ and $|CB|/|EB| = |AB|/|CB|$. (The symbol $|XY|$ means the length of the straight line from the point labeled X to the point labeled Y.) These equations algebraically translate into $|AC|^2 = |AE||AB|$ and $|CB|^2 = |EB||AB|$. Hence,

$$
\begin{aligned}
|AC|^2 + |CB|^2 &= |AE||AB| + |EB||AB| \\
&= (|AE| + |EB|)|AB| \\
&= |AB||AB| \\
&= |AB|^2
\end{aligned}
$$

4. He meant that the blobs were each scaled in proportion to the sides and hypotenuse—that is, if the magnitudes of the sides and hypotenuse were a, b, and c in order of size, the area of the middle-size blob equals k times the area of the smaller blob, where $k = b/a$. Similarly, the area of the largest blob is h times the area of the middle-size blob, where $h = c/b$.

5. Or did Roger place the proof in his translation of Jesus's words?

6. Their argument did not use forty-six propositions. But there was a catch. They used some intuitive notions that helped persuasion. For example, they used intuition about area to see that the area of the big triangle must be the sum of the areas of the smaller ones; after all, the two smaller triangles fit together to make the big one. They also "knew" the following syllogism:

    1. If two triangles are similar, they are proportional.
    2. If two triangles are proportional, they are the same except for size.
    3. If they are the same except for size, the ratios of their corresponding sides are equal.
    4. If the ratios of their corresponding sides are equal, $|AC|/|AE| = |AB|/|AC|$ and $|CB|/|EB| = |AB|/|CB|$.

    They then used some algebra to translate these equalities into this form:

    $$|AC|^2 + |CB|^2 = |AB|^2$$

    So they really did quite a bit of intuiting. In the end, they might have used many of the forty-six propositions.
7. This is an exaggeration. It's only 3,212 feet, but it's now known to be the tallest waterfall in the world. To get an idea of how high this is, multiply the height of Niagara Falls by 16.
8. A *regular* pentagon is a polygon with five equal sides.
9. Heath, *Euclid's Elements*, Vol. 4, 10.
10. Also best known for his Cours d'Analyse de l'École Polytechnique.

Chapter 3, *The Simple and Obvious Truth*

1. The renowned Cours d'Analyse de l'École Polytechnique.
2. One of the tools of Adobe Photoshop suggests the truth of the theorem. Dump paint onto any point in the illustration on the left, and Photoshop will paint either the inside or the

outside, leaving the illustration on the right. I thank my student Anthony Schein for this idea.

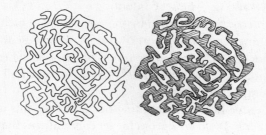

3. Gerald Holton, *Introduction to Concepts and Theories in Physical Science* (Reading, MA: Addison-Wesley, 1952), 69.

4. In sports, a referee's call is final, even when instant replay disagrees.

5. From the part of Leonard Shelby, the protagonist in *Memento*, a film by Christopher Nolan based on a short story by Jonathan Nolan.

6. David Hume, *An Inquiry Concerning Human Understanding*, ed. Charles W. Hendel (New York: Macmillan, 1955), 63.

7. If a number N is not prime, it must have a factor less than or equal to $\sqrt{N}$, so the only prime divisors to test are those less than or equal to $\sqrt{N}$. In the case of 8,191, one needs to test for primes less than or equal to 89. There are only twenty-three of them: 3, 5, 7, 11, 13, 17, 19, 23, 29, 31, 37, 41, 43, 47, 53, 59, 61, 67, 71, 73, 79, 83, and 89.

8. Andrew Weil, *Spontaneous Healing*, (New York: Alfred A. Knopf, 1995) 195.

9. Oswald Veblen, *American Mathematical Society Transactions* 6, 1905, 83–98.

Chapter 4, *What the Tortoise Said to Achilles*

1. L. Carroll, *Symbolic Logic* (Mineola, NY: Dover Publications, 1958).

2. Wittgenstein, *Tractatus Logico-Philosophicus* (Humanities Press, 1951), 55.

3. Because sometimes we might have statements of the form "If P and Q, then R," we need to have a containment metaphor for "P and Q." The set containing both P and Q is represented by *intersection* (denoted by the symbol $\cap$), which is the set of all things containing both P and Q. To represent "P or Q" by set inclusion, think of the *union* of two sets, where the *union* (denoted by the symbol $\cup$) means the set of all things containing either P or Q.

George Boole thought that he could take this idea one step further to build an algebra for set containment that mimics the traditional algebra of numbers. Could there be rules for manipulating sets as there are rules for manipulating numbers? It looks possible if we think of intersection as being a form of multiplication and union as a form of addition. Even the distributive laws hold. In algebra, the distributive law says this:

$$A \times (B + C) = A \times B + A \times C$$

Mimicking this law, we get this:

$$A \cap (B \cup C) = (A \cap B) \cup (A \cap C)$$

Boole thought he could build an axiomatic algebra for set theory—and, hence, for AND, OR, and NOT gates—and, hence, build laws of thought.

4. A different proof is given in Heath's *Euclid's Elements*, Book 10, Proposition 2.

5. For a truly fanciful account of what the tortoise said to Achilles, one should consult the full text, which is filled with puns and imaginative play.

6. My eighteen-month-old granddaughters say "mys" instead of the correct "mine," extending the logic of other possessive pronouns to the one that is inconsistent with the rule. The toddler hears *yours, hers, his, ours,* and *theirs* as possessives, so why isn't *mine mys*?

7. For extensive commentary on this point, see Heath, *Euclid's Elements*, 36–37, 195.

8. This statement, known as Playfair's axiom, is equivalent to Euclid's fifth postulate. It is attributed to the English mathematician John Playfair (1748–1819). Sir Thomas Heath's translation of Euclid's fifth postulate is this:

"That, if a straight line falling on two straight lines make the interior angles on the same side less than two right angles, the two straight lines, if produced indefinitely, meet on that side on which are the angles less than the two right angles." (See Euclid's *Elements*, trans. T. L. Heath (Mineola, NY: Dover Publications, 1956), 36–37, 202.

Chapter 5, *Legendre's Lament*

1. Quoted from Heath, *Euclid's Elements*, 119.
2. J. J. O'Connor and E. F. Robertson, http://www-history.mcs. st-and.ac.uk/history/Mathematicians/Legendre.html.
3. Of course, there are short straight lines that do not extend terribly far. Lines that extend for tens or hundreds of miles will have to bend to accommodate the spherelike shape of the planet.
4. Differential geometers use this word in a more general sense that is too advanced for the level of this book.
5. Angles are measured from the tangent lines to the geodesics.
6. You should convince yourself that the described circle is unique by sketching the possibilities.
7. The distance between two points A and B is measured as follows: Extend the geodesic through A and B to the points P and Q on the boundary of the Poincaré disk (in such a way that P is closer to B than to A). Then, letting $d(AB)$ denote the distance from A to B

$$d(AB) = \log\left|\frac{(\overline{AP})(\overline{BQ})}{(\overline{BP})(\overline{AQ})}\right|$$

Here, the horizontal bars represent the Euclidean length and the vertical bars represent absolute value. If either A or B is very close to the boundary of the Poincaré disk, the denominator is close to 0 and the numerator is not. So, the fraction is very large—hence, the log of the fraction is very large, and, hence the distance from A to B is very large. This definition of distance is what is expected from a distance measurement. With a bit of experience with logarithms, one can show that, for any three points A, B, and C, $d(AB) \leq d(AC) + d(CB)$.

8. Euclid may have been the principal author of *The Elements*, but his work should have acknowledged the contributions of many other mathematicians, including Anaxagoras, Archytas, Eudoxus, Menaechmus, Plilolaus, Theaetetus, and Theodorus. For example, it is believed that all of Book 5 (the theory of proportion) and the ideas behind Book 6 (the theory of exhaustion) were the work of Eudoxus, all of Book 8 was the work of Archytas, and the foundational ideas behind Book 10 (the theory of irrationals), as well as some of the ideas behind the theory of the five regular solids in Book 13, were the work of Theaetetus.

9. Tobias Dantzig, *Number: The Language of Science* (New York: Doubleday, 1954), 105.

Chapter 6, *Evan's Insight*

1. Isak Dinesen, *Out of Africa* (New York: Random House, 1992).
2. From *Phibebus and Epinomis*, trans. (with an introduction) A. E. Taylor, ed. R. Klibansky (New York: Barnes & Noble Books, 1972), 977c.
3. Ibid., 978b.
4. D. E. Smith and J. Ginsburg, *The World of Mathematics, From Numbers to Numerals and From Numerals to Computation, In The World of Mathematics*, Vol. 1, ed. J. R. Newman (New York: Simon and Schuster, 1956), 442–463.

5. Remember that the symbol 10^k means 10 multiplied by itself k times. The result is called *a power of 10*.

6. The symbol 2^k means 2 multiplied by itself k times; also recall that $2^0 = 1$.

7. I use these examples because our mature understanding of three-ness originates with very young associations with three things.

8. I now have quite a few resources for teaching these elementary topics, the best of which is the recent and magnificent book by Liping Ma, *Knowing and Teaching Elementary Mathematics* (Mahwah, NJ: Lawrence Erlbaum Associates, 1999). It is a comparison of teachers' understanding of elementary mathematics in China and the United States.

9. J. Wassmann and P. R. Dasen, "Yupno Number System and Counting," *Journal of Cross-Cultural Psychology* 25 (1994): 78–94.

10. D.E. Smith, *History of Mathematics*, (New York: Dover, 1958).

11. For further details and an interesting excursion into how finger counting was used by traders in different cultures, see Karl Menninger, *Number Words and Number Symbols: A Cultural History of Numbers* (Mineola, NY: Dover Publications, 1992), 201–220.

12. This works because $ab = (a - 5 + b - 5) \times 10 + (10 - a) \times (10 - b)$.

13. The reason for this comes from the algebraic identity $ab = ((a - 10) + (b - 10))) \times 10 + (a - 10) \times (b - 10) + 100$. A similar method works for any two numbers, but it gets more complicated because more than five fingers might have to be raised on each hand. The idea is to make use of the formula $ab = ((a - c) + (b - c)) c + (a - c) (b - c) + c^2$, where c is the amount the numbers should be reduced to bring the multiplication down to a manageable number. Unfortunately, larger numbers require knowing how to square c.

14. D. E. Heath, *History of Mathematics* (Mineola, NY: Dover Publications, 1953), 119–20.

15. B. Butterworth, *What Counts: How Every Brain Is Hardwired for Math* (New York: Free Press, 1999).

16. W. Penfield and T. Rasmussen, *The Cerebral Cortex of Man* (New York: The Macmillan Company, 1952).

17. David Hume, *A Treatise of Human Nature* (New York: Penguin, 1985), 58.

18. Francis Galton, *Inquiries into Human Faculty,* (University Press of the Pacific, 2003), 146.

19. Jacques Hadamard, *Psychology of Invention in the Mathematical Field* (Mineola, NY: Dover Publications, 1990).

20. Evan's scheme for deciding whether a point is on the inside or outside does not prove that the curve divides the plane into two distinct regions. His scheme must be modified to take into account unexpected anomalies. It would also have to show that any two points on the inside can be connected by a curve that does not cross the original curve, and any two points on the outside can do the same. The point here is that Evan's scheme was the kernel of the right idea that would eventually lead to an airtight proof.

Chapter 7, *Encounters on the Aegean*

1. This was how Archimedes calculated that π is between 2130/71 and 21/7.

2. There are many variations on this type of construction. If the starting figure is a semicircle, for instance, with each new construction doubling the number of semicircles while keeping the same total length, then one can show that $\pi = 2$.

Chapter 8, *Zindo the Trojan Superman*

1. Note that this means that there are many points on the sawtooth figure that are not on the straight line. For example, a point at distance $1/\sqrt{2}$ units from the bottom is not on both figures.

2. This means that there are many points on the infinite-sided polygon that are not on the circle. For example, a point at distance $\pi/(3\sqrt{2})$ units from the top of the circle is not on both figures.

3. Perimeter.

4. Edith Hamilton and Huntington Cairnes, *The Collected Dialogues of Plato*, trans: Francis Macdonald Cornford (Princeton, New Jersey: Princeton University Press, 1961), 921–922, 127a–d.

5. The Hilbert quote is a translation from the German found in Stephen Kleene, *Introduction to Metamathematics* (Princeton, New Jersey: Van Nostrand, 1962), 54–55.

Chapter 9, *Finding Pegasus*

1. In other words, the number whose square is 23, or that number which gives 23 when multiplied by itself.

2. The actual number π cannot be written as a decimal expansion with a finite number of digits.

3. I could not find anything in the usual Zeno fragments that uses Achilles and the tortoise for this argument. It seems to have been made up by modern authors to color the story. The main sources for Zeno's arguments are in Aristotle's *Physics*, Diogenes Laertius's *Lives of the Philosophers*, Simplicius's *Commentary on the Physics*, and Plato's *Parmenides*. None of these sources talks about a tortoise.

4. A number is called *rational* if it can be represented by a fraction p/q, where both p and q are positive or negative whole numbers, and (of course) q is not equal to 0. *Irrational*, then, means "not rational." All we have done is to give an alternative name to any number that has a certain specified property. Rather than use a lengthy description such as "a number that can be represented by a fraction p/q, where both p and q are whole numbers, and q is not equal to 0" every time we want to refer to such a number, we prefer to give it a brief sign marked by the word *rational*. The name chosen, *rational*, is old and comes from a time when such numbers were thought of as rational (reasonable—coming from reason), as opposed to irrational (unreasonable). In this age, even Jeremy would agree that there is nothing unreasonable about $\sqrt{2}$, and yet we say that it is irrational.

5. We find this proof in Aristotle's *Prior Analytics*, but also in Heath's introductory note to Euclid's *Elements*, Book 10.

6. An irrational number is a number that cannot be written as a fraction with both its numerator and its denominator being positive or negative whole numbers. One way to think of an irrational number is to think of a never-ending decimal expansion that has no repeating pattern.

7. The proof is given on p. 60–61.

8. Of course, this assumes that you accept two principles: that if the statement's negation is false, the statement is true; and that the statement is either true or false, *the law of the excluded middle.*

9. Equations with no solutions often lead us to places that are more interesting than do those with solutions. We now know that the integer equation $x^n + y^n = z^n$ has no solutions for n greater than 2, but it took 350 years of extraordinarily interesting developments in number theory to prove it.

10. The discovery of irrationals must have been disturbing in Pythagoras's time and might even have obstructed progress in proving some of the early theorems of geometry, but the crisis was resolved by the ingenious idea of Eudoxus to use proportion to skirt the assumption that a number could be irrational.

11. Lewis Carroll, *Through the Looking Glass and What Alice Found There* (New York: Random House, 1946).

12. *Rene Descartes: Meditations on First Philosophy*, ed. John Cottingham (Cambridge University Press, 1996), 16–17.

13. For example, suppose we have the number 1.41414141..., which has a repeated decimal expansion.

Label that number as x.
Multiply x by 100 to get $100x = 14.14141414....$
Subtract by x to get $99x = 14$.
Solve for x to get $x = 14/99$.

You can play the same trick for any decimal number that has a finite number of digits or a repeating pattern.

14. Suppose we give a name to things that are never-ending sequences of digits that have no repeating pattern: Call them *irrational objects*. Then ask, "Are irrational objects numbers?" We surely would like them to be; after all, we just claimed that √2 is an irrational object, and √2 represents the measurement of the side of a square of length 1. So, can we simply admit these objects to the number club? Perhaps. We can if their existence as numbers will not contradict anything that we already know about our old numbers. You can check that a number cannot be both an irrational and a rational object.

15. For obvious reasons, the representation of √2 displayed here is called a *continued fraction*. The right side is just 1 + the fraction 1 over 2 + the fraction 1 over 2 + the fraction 1 over 2, and so on.

16. To see this, just let

$$x = 1 + \cfrac{1}{2 + \cfrac{1}{2 + \cfrac{1}{2 + \cfrac{1}{2 + 1_{\cdot_{\cdot_{\cdot}}}}}}}$$

Then

$$x = 1 + \frac{1}{1 + x}$$

A bit of algebra gets this last equation to become $x^2 = 2$. According to A. E. Taylor, the great early-twentieth-century Plato scholar, referring to the method of computing √2 by stopping partial fractions at some finite length, "[T]his method of finding the value of what we call √2 was pretty certainly known to Plato." The reference here is from A. E. Taylor's *Plato: The Man and His Work* (New York: Meridian Books, 1960), 510.

17. This is true. In general,

$$n = 1 + \cfrac{(n-1)n}{1 + \cfrac{(n-1)n}{1 + \cfrac{(n-1)n}{1 + \cfrac{(n-1)n}{1 + \cfrac{(n-1)n}{\ddots}}}}}$$

But this form of partial fraction is not conventional. *Simple partial fractions* are defined as partial fractions of this form:

$$x = a_1 + \cfrac{1}{a_2 + \cfrac{1}{a_3 + \cfrac{1}{a_4 + \cfrac{1}{a_5 + 1 + \ddots}}}}$$

Here, a_1, a_2, a_3, and so on are integers. Note that the numerators of the fractions all have the value 1. So, integers cannot be written as simple partial fractions.

Chapter 10, *Some Things Never End*

1. Simone Weil, *Notebooks*, trans. Arthur Wills, Routledge & Kegan Paul, p. 424.
2. This assumes that the intervals are all greater than some fixed real number.
3. David Hume, *A Treatise of Human Nature* (New York: Penguin, 1985), 76.
4. René Descartes, *A Discourse on Method,* trans. John Veitch, J.M. Dent, (New York, 1937).
5. Ibid.
6. *Guinness Book of World Records.*

7. *Induction Hypothesis:* Suppose that the following two conditions are satisfied:

    1. The statement "If $S(n)$ is true, then $S(n+1)$ is true" is true for every n.

    2. $S(1)$ is true.

Then $S(n)$ is true for all whole numbers.

If you accept this axiom, then we can easily prove that

$$1 + 2 + 3 + \ldots + n = \frac{n(n+1)}{2}$$

8. Proof 1 (a proof that relies on a concept of an infinite construction)

Surely $S(1)$ is true because $S(1)$ states that $1(1+1)/2$. (The first domino falls.)

We now show that the truth of $S(n)$ implies the truth of $S(n+1)$. Start with $S(n)$, which states that

$$1 + 2 + 3 + \ldots + n = \frac{n(n+1)}{2}$$

We are assuming that $S(n)$ is true. (In other words, the nth domino falls.)

Add $(n+1)$ to both sides to get the following new equation that is also true:

$$1 + 2 + 3 + \ldots + n + (n+1) = \frac{n(n+1)}{2} + (n+1)$$

Use the rules of algebra to simplify the right side to

$$(n+1)\frac{(n+1)+1}{2}$$

Now you have the following equation, which is also true:

$$1 + 2 + 3 + \ldots + n + (n+1) = (n+1)\frac{(n+1)+1}{2}$$

But this last equation is simply $S(n+1)$. (The nth domino knocked down the $(n+1)$st domino.)

So we have shown that $S(n+1)$ must be true every time $S(n)$ is true for all whole numbers.

9. Proof 2 (A finite algebraic proof)

One can simply write the sum backward, add the two numbers together, and notice that the result will be n copies of $(n+1)$.

$$
\begin{array}{l}
1 \quad + \quad 2 \quad + \quad 3 \quad + \ldots + \quad n, \\
\underline{n \quad + (n{-}1) + (n{-}2) + \ldots + \quad 1} \qquad \text{(written backward)} \\
(n{+}1) + (n{+}1) + (n{+}1) + \ldots + (n{+}1) \qquad \text{(added together)}
\end{array}
$$
(There are n copies of $[n{+}1]$.)

The sums $1 + 2 + 3 + \ldots + n$ and $n + (n{-}1)+(n{-}2) + \ldots + 1$ are the same number. So, the n copies of $(n+1)$ must equal twice the sum $1 + 2 + 3 + \ldots + n$, which means that

$$
1 + 2 + 3 + \ldots + n = \frac{n(n+1)}{2}
$$

Chapter 11, *All Else Is the Work of Man*

1. Note that the word *number* is italicized. Because there are infinitely many points in both the square and one of its sides, we should define what we mean by the word *number* when used in this context.

2. For example, to solve the quadratic equation $t^2 + 2t - 8 = 0$, start with a square of side t, add two rectangles of width 1 and length t to one side of the square, flip one of the rectangles to the other side of the square, and add what you need to make the figure you have into a complete square.

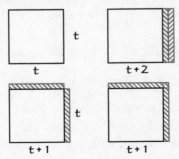

The algebraic analog of what you have done is this: Start with $t^2 + 2t - 8 = 0$. Notice that $t^2 + 2t$ represents the area of the second and third figures. Add a square of side 1 to the upper right hand corner to get the area of the figure to the right. You now have a perfect square with sides equal to $t + 1$. To compensate for the added square of side 2, subtract 1^2 from the left side of the equation. Then, from the geometry, notice that $t^2 + 2t - 8$ is the same as $(t+1)^2 - 9$. So we can interpret the equation $t^2 + 2t - 8 = 0$ geometrically to mean that the perfect square $(t+1)^2$ is the same as the perfect square 9. This means that $(t+1)^2 = 9$ or that $(t+1) = \pm 3$, which, in turn, means that $t = -4$ or $t = 2$.

3. Recall that a rational is a number that is expressible as a ratio of two integers.

4. A polynomial equation with integer coefficients is an equation of the form $A_n x^n + A_{n-1} x^{n-1} + \ldots + A_1 x + A_0 = 0$, where the terms A_n, A_{n-1}, ... A_1, A_0 are fixed integers (whole numbers) and n (called the *degree* of the polynomial) is any positive integer. For example, $x^2 + 2x - 8 = 0$ is a polynomial equation in which $n = 2, A_2 = 1, A_1 = 2$, and $A_0 = -8$. For π to be constructible, it must be the solution to a polynomial of degree 2^k for some integer k.

5. This is because constructions can be modeled only by polynomial equations with integer coefficients.

6. It is a subset, not necessarily a member.

7. Simone Weil, *The Notebooks of Simone Weil*, Vol. 2, trans. Arthur Wills (London: Routledge & Kegan Paul, 1956), 362.

8. *Dictionary of Scientific Biography* (New York: Charles Scribner's Sons, 1970–1990).

9. Jorge Luis Borges, *Other Inquisitions*, translated by Ruth L. C. Simms (Austin, TX: University of Texas Press, 1975), 179.

10. The origins of this discovery are debatable. Some historians believe it was Pythagoras.

Chapter 12, *A Fistful of Chips*

1. Charles de la Vallee Poussin independently proved the prime number theorem at the same time as Hadamard.
2. Without such consideration, the face with one spot has more material than one with a higher number of spots. Take a penny, for example. You might expect a flip of a penny to come up heads just as often as tails. In fact, the slight difference in weight between the two sides makes the penny biased toward heads. Try flipping a penny to see if this is true.

Chapter 13, *Who's Got a Royal Flush?*

1. Anyone can guess that pollution (from busses or coal-burning power plants) is somewhat responsible for respiratory diseases, but how could one know for sure? One can imagine how important the answer is for public health information and the energy policy legislation. Johns Hopkins researchers are attempting to answer that question by correlating pollution statistics with mortality rates by examining data from a colossal Medicare databank (40 million subscribers) and daily pollution records. The Johns Hopkins study is scheduled to complete by 2006.
2. This story comes from an essay by English statistician and geneticist Sir Ronald Fisher, printed in *The World of Mathematics*, Vol. 3, ed. James E. Newman (New York: Simon and Schuster, 1956).
3. Fisher's essay is really meant to be about the design of experiments and the concern over subjective error, but the professor was using the story to point to the connection between mathematics and experiment.
4. This has to be qualified. For this product to be true, the two events must be independent—that is, the occurrence of one does not affect the occurrence of the other. You can check this by experiment. Try tossing a coin and a die, and ask for the probability of tossing a head and an even number. There is

only one head on the coin and there are three even-numbered sides of the die. The one head can happen with any one of the three even-numbered sides. You have twelve possible outcomes: heads with six possible faces and tails with six possible faces. But there are only three successful outcomes; hence, the probability of tossing a head and an even number is 1/4. Now look at the product of the probability of tossing a head and the probability of rolling an even number. You get $1/2 \times 1/2$.

5. In general, the symbol $C(n,k)$ is a symbol in combinatorial analysis that stands for the number of ways of arranging combinations of n objects taken k at a time. For example, $C(3,2) = 3$ and three objects—A, B, and C—can be arranged in three ways as AB, AC, and BC. $C(4,2) = 6$. Four objects—A, B, C, and D—can be arranged in six ways as AB, AC, AD, BC, BD, and CD. It turns out to be n!/k!(n–k!). The symbol $C(20,k)$ stands for 20!/k!(20–k)!. The symbol $k!$ stands for the product of all numbers from 1 to k. Therefore, $(20 - k)!$ stands for the product of all numbers from 1 to $(20 - k)$. For example, if $k = 8$, then $(20 - k) = 12$ and we would have $8! = 1\times2\times3\times4\times5\times6\times7\times8 = 40{,}320$, and $12! = 1\times2\times3\times4\times5\times6\times7\times8\times9\times10\times11\times12 = 479{,}001{,}600$. The number of different ways eight blacks can be chosen from a bowl by taking twenty at a time is 20!/(8!12!) = 125,970.

6. This, too, must be qualified. For this sum to be true, the two events must be mutually exclusive—that is, they cannot both occur at the same time. You can check this by experiment also. Try rolling a die and ask for the probability of rolling a number larger than a 4. There are only two favorable outcomes—a 5 or a 6. You have six possible outcomes—the six faces. Because there are two successful outcomes, the probability of rolling a 5 or a 6 is 2/6 or 1/3. Now look at the sum of the probability of rolling a 5 and the probability of rolling a 6. You get 1/6 + 1/6.

7. Notice that the sum of these probabilities is very close to 1. (It's actually 1.006, but the 0.006 is due to round-off error.)

8. The actual table (to nine decimal places) is shown here:

Number of Black Chips in Sample	Probability of Picking k Black Chips from Twenty
0	0.000797922
1	0.006839337
2	0.027845872
3	0.071603672
4	0.130420974
5	0.178863050
6	0.191638982
7	0.164261985
8	0.114396739
9	0.071578259
10	0.030817080
11	0.012006654
12	0.003859281
13	0.001017832
14	0.000218106
15	0.000037389
16	0.000005007
17	0.000000504
18	0.000000036
19	0.000000001
20	0.00000000003

9. To calculate this particular standard deviation, just take the square root of the product of the mean and the probability of picking a white chip.

10. Ninety-five percent of the area under the curve is between two standard deviations on either side of the mean.

Chapter 14, *Boxcars and Snake Eyes*

1. *The Iliad of Homer*, Book 15, trans. Richard Lattimore (New York: HarperCollins, 1974).

2. This was first published in 1663.

3. The typical board-game die has its dots gouged from the sides of a cube. Each gouge is as deep as the next, so the side with six gouges is lighter than the side with one gouge. Such a die is dishonest because it favors heavier sides. To make an honest die, material gouged from one side should weigh the same as the material gouged from any other side. The paint to make the dots should also be weighed and balanced.

4. It is also possible that the penny will land on its edge for the next hundred flips.

5. The law of large numbers is deducible from axioms of probability. Alas, its proof is extremely long and is beyond the scope of this book. For more information, see W. Feller, *An Introduction to Probability Theory and Its Applications*, Vol. 1, 3rd ed. (New York: Wiley, 1971).

6. There are two laws of large numbers, the strong and the weak. Here we are using what is called the *strong law of large numbers*, which says that, with probability 1 (certainty), the sample mean approaches the true mean as n (the sample size) grows large. The weak law says that, with some probability $P(n)$, which approaches 1 as n grows large, the sample mean approaches the true mean as n grows large.

7. Hesoid, *Theogony*, trans. Norman Brown. (New York: Liberal Arts Press, 1953) 56.

8. We are using here a limited definition of plausible reasoning. Plausible reasoning usually means far more than just the kind of reasoning involved in statistical hypothesis testing. Often it means reasoning by analogy: Something is true because something similar is true. We have seen this sort of reasoning in corollaries of the Pythagorean theorem: To find the square of the diagonal of a rectangle of sides a and b, just take the sum of the squares of a and b. So, it would be plausible to guess that to get the square of the diagonal of a rectangular solid of sides a, b, and c, just take the sum of the squares of a, b, and c. Indeed, that guess would be correct. It is also common for mathematicians to guess that a generalization from a specific case might be true. Though it often is not, there can

be enough plausibility in the generalization to warrant investigation, and the process of generalization can entice broader understanding of specifics. I have avoided these topics by choice. For stimulating reading on the broader notions of plausible reasoning, I refer my readers to George Polya's *Mathematics and Plausible Reasoning,* Vol. 1 and 2 (Princeton, NJ: Princeton University Press, 1990).

9. David Hume and Tom L. Beauchamp, *An Energy Concerning Human Understanding* (Oxford University Press, 1999).

10. J.M. Keynes, *A Treatise on Probability* (Dover Publications, 2004).

11. See Mark Kac, "Probability," *Scientific American* (September 1964), 92–108.

Chapter 15, *Anna's Accusation*

1. Rolling 7 or 11 on the first shoot wins. A 2, 3, or 12 on the first throw loses. A throw of either 4, 5, 6, 8, 9, or 10 wins, if it can be duplicated before a 7. The interesting thing here is that the mathematical chances of each of these events must have been known before the game was invented.

2. To solve the equation $1 - (35/36)^n = 1/2$, subtract 1/2 from each side to get $1/2 - (35/36)^n = 0$ or $1/2 = (35/36)^n$. Take the natural logarithm of both sides to get $Ln 1/2 = nLn(35/36)$. Isolate n by dividing both sides by $Ln(35/36)$ to get $n = Ln(1/2)/Ln(35/36) \approx 24.6$. If n is to be an integer that does the trick, it must be 25.

3. $1 - \left(\dfrac{35}{36}\right)^{24} < \dfrac{1}{2}$, but $1 - \left(\dfrac{35}{36}\right)^{25} > \dfrac{1}{2}$.

4. J. Boyer, "DNA on Trial," *The New Yorker*, Jan. 17, 2000.

5. A better experiment—but a far more complex one—would involve two or more groups of facilitators and two or more groups of children of similar ages and the same gender. Then the pictures could be switched between the groups to

eliminate any problems that would result from the interpretations of pictures (the concepts they represent) and the way the pictures are chosen.

6. D. L. Wheeler, et al., "An Experimental Assessment of Facilitated Communication," *Mental Retardation* 31 (1993), 49–60.

7. Natalie Russo and the Commission on Quality of Care, "Facilitated Communication," http://www.cqc.state.ny.us/fcnatal.htm.

8. American Psychological Association, "Resolution on Facilitated Communication" (14 August 1994), reprinted at http://web.syr.edu/~thefci/apafc.htm.

9. Commission on Quality of Care for the Mentally Disabled, "Facilitated Communication," http://www.cqc.state.ny.us/fchot.htm.

10. Vermont Facilitated Communication Network, http://www.uvm.edu/~uapvt/faccom.html.

11. The Galton Board is also called the quincunx because of the quincunical arrangement of pins—the same arrangement as five spots on a die.

12. I am grateful to Susan Holmes at Stanford for referring me to the best demonstration of the Galton Board, at http://www.jcu.edu/math/isep/Quincunx/Quincunx.html.

13. Mark Kac, "Probability," *Scientific American* (September 1964), 92–108.

Chapter 16, *Dr. Mortimer, I Presume?*

1. Francis Bacon, *The New Organon*, ed. Fulton H. Anderson (New York: Library of Liberal Arts, 1960), 41.

2. Leo Tolstoy, *War and Peace*, trans. Constance Garnett (New York: Modern Library, 1940), 768.

3. Discovered by Michael Shafer on November 17, 2003.

4. A proper divisor of an integer n is an integer less than n that divides n.

5. Euclid, *Elements*, Book 9, Proposition 36.

Conclusion

1. Roger felt that the number represented the paraboloidal shape of the spinning potion. The shape of a paraboloid is completely determined by a single number that tells how sharply the bowl-like shape cups upward. If you want to describe where it is or how it is oriented in space, you will need more than one number. But the shape is totally described by just one number. Roger was identifying paraboloids with numbers—one number for each paraboloid and one paraboloid for each number. It seemed to be what the mambo was saying. Forty years ago, I thought about this and came to the same conclusion. A paraboloid is just a spinning parabola, which is uniquely given by a cone, which, in turn, is given by an angle (the angle of the cone)—a single number. Euclid understood that two points determine a unique line, but he never mentioned that a circle and a point determine a cone, which, in turn, determines a parabola. To get a different parabola, you would cut a different cone, and each cone depends on the angle that it makes with the vertical axis. Revolve a line around the circle, keeping it fixed at the point P, and you will generate a cone.

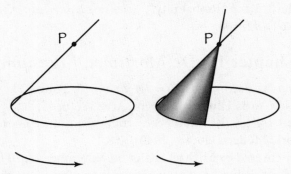

Cut the cone with a plane in a direction parallel to the generating line and the points in contact with the plane form a parabola. Now think about this: does the parabola that you get depend on the cutting plane, assuming that any cutting plane is parallel to the generating line? For example, if you cut the cone with a plane that is lower, will it cut out a different parabola? No. Why? Because the cone is independent of scale. If you enlarge the cone at its vertext then you see exactly the same cone.

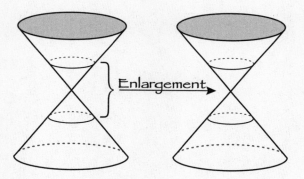

This means that the form of a cut by a plane will not depend on where the cut is made. But if this is true, the cone determines a unique parabola! In turn, it means that a point and a circle determine a unique parabola. The cone is completely determined by the angle that it makes with its vertical axis.

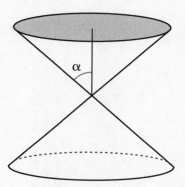

Appendix 1, *Proof That All Triangles Are Isosceles*

1. From *Unknown Lewis Carroll*, ed. Stuart Dodson Collingwood (New York: Dover Publications, 1961), 264–265.

Further Reading

Mathematics Accessible to Readers with Little or No Background

Aczel, Amir. *The Mystery of the Aleph: Mathematics, the Kabbalah, and the Search for Infinity*. New York: Washington Square Press, 2000.

> A gripping book tracing the development of infinity through the life of Cantor. Many of the concepts brought up in Part 2 of *Euclid in the Rainforest* are explained in this book.

Artmann, Benno. *Euclid: The Creation of Mathematics*. New York: Springer, 1999.

> A clear, friendly, and nicely annotated introduction to Euclidean geometry.

Bell, E. T. *Men of Mathematics*. New York: Simon & Schuster, 1937.

> A book of interesting anecdotes surrounding mathematicians up to the nineteenth century. This is easy reading, with few formulas and very little mathematics.

Berlinghoff, William, and Fernando Gouvêa. *Math Through the Ages: A Gentle History for Teachers and Others*. Farmington, Maine: Oxton House Publishers, 2002.

> A very well organized and friendly book. It is broken into two parts: a general overview and appendixes "sketches" providing more detail. The beauty of this book is in its exposition: The

overview does not interrupt the flow with details that are continuously linked to the sketches. One is reminded of Internet
reading with highlighted words signaling links to greater
detail.

Bolzano, Bernard. *Paradoxes of the Infinite*. Translated by Dr. Fr.
Prihonsky. London: Routledge and Kegan Paul, 1950.
     I include this reference for the historical introduction by Don-
ald Steele, which amounts to almost half the book. The language is archaic, but the historical details create a strong
picture of a period when infinity was actively being studied.
Bolzano's book has hundreds of highly interesting paradoxes
of the infinite that are quite accessible to anyone without a
math background who wants to wade through the archaic language.

Bradis, V. M., V. L. Minkovskii, and A. K. Kharcheva. *Lapses in Mathematical Reasoning*. Mineola, New York: Dover Publications, 1999.
     This book was designed to teach mathematical thinking to
high school students in Soviet Russia. These carefully chosen
fallacies are carefully analyzed and clearly exposed.

Dantzig, Tobias. *Number: The Language of Science*. 4th ed. Garden
City, New York: Doubleday, 1956. A classic.
     A very readable and clear introduction to mathematics. The
first two thirds of this marvelous book brings readers from the
earliest beginnings of mathematics to nineteenth-century
exploration and understanding of the infinite.

Dehaene, Stanislas. *The Number Sense: How the Mind Creates Mathematics*. New York: Oxford University Press, 1997.
     A readable, interesting account of how humans (and animals)
think about mathematics. In particular, Chapter 2 is about the
number sense of newborns, and Chapter 4 is about how
humans conceptualize numbers.

Dunham, William. *Journey Through Genius: The Great Theorems of Mathematics.* New York: John Wiley & Sons, 1990.

> A nicely organized book that touches on many of the mathematical ideas talked about in parts 1 and 2 of *Euclid in the Rainforest.*

Gamow, George. *One Two Three...Infinity: Facts and Speculations of Science.* New York: Viking, 1961.

> Start reading, and you will find that it is hard to stop. Though this book was written in the middle of the last century, it still has a magnificent freshness that only a good storyteller like Gamow can tell. The book makes surprising connections between eclectic branches of mathematics and science. There is plenty here to introduce you to infinity.

Hogben, Lancelot. *Mathematics for the Million.* New York: W. W. Norton, 1937.

> Read Chapter 4 of this book, "Euclid Without Tears," to see that it is really possible.

Kaplan, Robert, and Ellen Kaplan. *The Art of the Infinite: The Pleasures of Mathematics.* New York: Oxford University Press, 2003.

> Like Kaplan's other book, *The Nothing That Is*, this witty book is a page turner. Kaplan's poetic writing style brings a unique pleasure to novices reading about mathematics. It is an excellent elementary compressive treatment of the infinite.

Lakoff, George, and Rafael Nunez. *Where Mathematics Comes From: How the Embodied Mind Brings Mathematics into Being.* New York: Basic Books, 2000.

Mazur, Barry. *Imagining Numbers (Particularly the Square Root of Minus Fifteen).* New York: Farrar Straus Giroux, 2003.

> An easily readable, enlightening account of imagination in poetry and mathematics. Chapter 2, "Square Roots and the Imagination," and Chapter 3, "Looking at Numbers," have particular relevance to several points discussed in *Euclid in the Rainforest.*

Moore, David. *Statistics: Concepts and Controversies*. San Francisco: Freeman, 1979.
> This is a remarkably readable introduction to statistics written in the form of a textbook. Moore carefully exposes the concepts without using much mathematics.

Stewart, Ian. *Concepts of Modern Mathematics*. Mineola, New York: Dover, 1995.
> This is precisely about what the title says it's about. If you've ever read other books by this author, you will know that the reading will be clear, concise, accurate, current, and lucid.

Stewart, Ian. *Life's Other Secret: The New Mathematics of the Living World*. New York: John Wiley & Sons, 1998.
> A fascinating, lively book filled with up-to-date applications of mathematics to the physical world.

Weaver, Warren. *Lady Luck: The Theory of Probability*. Garden City, New York: Anchor Books, 1963.
> An extremely math-friendly introduction to probability through games and chance occurrences of everyday life. See Chapters IV, X, and XII for well-explained background material to Part 3 of *Euclid in the Rainforest*. See Chapter XI for more on the law of large numbers.

Whitehead, Alfred North. *An Introduction to Mathematics*. New York: Henry Holt & Company, 1939.
> This small book is a collection of important basic ideas necessary to learning mathematics. It is a bit out of date but is still very readable.

Mathematics Accessible to Readers with a Solid High School Background

Courant, Richard, and Herbert Robbins. *What Is Mathematics?* New York: Oxford University Press, 1963.
> This book entices curious readers to pursue a deeper understanding of the wonderful collection of interesting topics it explores from many branches of mathematics.

Ekeland, Ivar. *Mathematics and the Unexpected.* Chicago: University of Chicago Press, 1988.

> Chapter 2 of this book discusses Poincaré's general theorem of dynamical systems as it relates to some of the apparent paradoxes surrounding properties of gases discussed in Chapter 15 of *Euclid in the Rainforest.*

Euclid. *The Elements.* Translated and edited by Sir Thomas Heath. 3rd ed. revised, with additions. Mineola, New York: Dover Publications, 1956.

> The commentaries in this edition of Euclid's great work are as interesting as Euclid's proofs. See the commentaries following Proposition 47 for various details on Western and Eastern versions of the proof of the Pythagorean theorem.

Greenberg, Marvin Jay. *Euclidean and Non-Euclidean Geometries: Development and History.* 3rd ed. New York: W. H. Freeman, 1993.

> If you want to learn about non-Euclidean geometries, this is the book. It is clearly written and filled with reasonable exercises designed to give the reader an intuitive sense of non-Euclidean worlds.

Lavine, Shaugham. *Understanding the Infinite.* Cambridge: Harvard University Press, 1998.

> In this book, Lavine gives original ideas surrounding the philosophy and history of infinity. Parts are accessible to the general reader, but much of this book is addressed to mathematically sophisticated audience.

Polya, George. *Mathematics and Plausible Reasoning.* Vols. 1 and 2. Princeton, New Jersey: Princeton University Press, 1990.

> Polya is the master of exposition. The topics in this book are not always elementary, but Polya, the world-renowned master of exposition, has a style that makes you believe that they are.

Ruelle, David. *Chance and Chaos.* Princeton, New Jersey: Princeton University Press, 1991.

> A fascinating, witty, amazingly readable book dealing with current issues surrounding chance and chaos.

Singh, Jagjit. *Great Ideas of Modern Mathematics: Their Nature and Use.* Mineola, New York: Dover Publications, 1959.

Singh, Jagjit. *Mathematical Ideas.* Mineola, New York: Dover Publications, 1959.
    Chapter 5 contains a wonderful exposition of Zeno and infinity. This is also a good source for learning about sets (see Chapter 5) and probability (Chapter 9).

Stewart, Ian, and David Tall. *The Foundations of Mathematics.* Oxford: Oxford University Press, 1977.
    Clearly written and true to its title, this book is for readers who are seriously interested in exploring higher mathematics.

Philosophy of Mathematics

Black, Max. *The Nature of Mathematics: A Critical Survey.* London: Routledge & Kegan Paul, 1933.

Bochner, Salomon. *The Role of Mathematics in the Rise of Science.* Princeton, New Jersey: Princeton University Press, 1966.

Buchanan, Scott. *Truth in the Sciences.* Charlottesville, Virginia: University Press of Virginia, 1972.

Dauben, Joseph. *Georg Cantor: His Mathematics and Philosophy of the Infinite.* Princeton: Princeton University Press, 1990.
    This is a comprehensive work on the philosophy of the infinite. It requires considerable math background.

Goodstein, R. L. *Essays in the Philosophy of Mathematics.* Leicester: Leicester University Press, 1965.
    This is a compilation of very readable essays reprinted from several respected journals by Reuben Goodstein, a prolific writer known for his clear, expository style. For readers who want to quickly understand the notions of proof and the axiomatic method without much work, the essays in this book are excellent.

Hadamard, Jacques. *An Essay on the Psychology of Invention in the Mathematical Field.* Mineola, New York: Dover Publications, 1954.
> This is one of the rare readable books on how the mind invents mathematics written from the point of view of a mathematician. For others, see the Poincaré references.

Holton, Gerald. *Introduction to Concepts and Theories in Physical Science.* Reading, Massachusetts: Addison-Wesley, 1952.

Hume, David. *An Inquiry Concerning Human Understanding.* Edited by Charles W. Hendel. New York: Macmillan, 1955.
> Hume is one of those eighteenth-century philosophers who can communicate well with intelligent readers. Sections 4, 5, and 6 are of particular interest here for the clear way he looks at human reasoning and probability.

Menninger, Karl. *Number Words and Number Symbols: A Cultural History of Numbers.* Translated by Paul Broneer. Mineola, New York: Dover Publications, 1992.
> This book is extremely comprehensive and filled with the details of cultural history of numbers. Read any section and become absorbed in wonderful facts about the evolution of number writing, symbols, and cultural notions of counting. The book is filled with photographs and drawings of ancient counting, calculating, and measuring artifacts.

Poincaré, Henri. *Science and Hypothesis.* Mineola, New York: Dover Publications, 1952.

Rucker, Rudy. *Infinity and the Mind.* Princeton, New Jersey: Princeton University Press, 1995.
> I have taught freshmen from this book with great success. It was a favorite in a collection of readings for a class on infinity. Clarifies Cantor's arguments and explores infinity in all its forms from different points of view. This book includes a very clear exposition of Gödel's incompleteness theorems.

Russell, Bertrand. *Introduction to Mathematical Philosophy.* London: George Allen and Unwin, 1919.
> This is an amazingly clear exposition of the foundations of natural numbers. In just three short chapters, Russell, in his inimitable style, gets to infinity and induction.

History of Mathematics

Allman, George. *Greek Geometry from Thales to Euclid.* New York: Arno Press, 1976.
> This nineteenth-century work complements Proclus's book, which, in turn, summarized a lost history of geometry written by Eudemus in the third century B.C. It contains a comprehensive, well-written commentary on fifth-century B.C. mathematics surrounding the Pythagorean mathematics.

Boyer, Carl. *A History of Mathematics.* New York: Wiley, 1968.
> This is one of the standard references for mathematicians on the history of mathematics. Boyer gives a clear understanding of the ideas contributed by the Pythagoreans, Zeno, Eudoxus, and Plato.

Dijksterhuis, E. J. *Archimedes.* Princeton, New Jersey: Princeton University Press, 1987.
> This is a magnificent book of commentaries on Archimedes's life and work. Though no background beyond a familiarity with Euclid is necessary, reading this book requires considerable thought and work, the kind most mathematics enthusiasts love to do.

Fowler, David. *The Mathematics of Plato's Academy.* London: Oxford University Press, 1987.
> A refutation of the traditionally accepted story that Greeks abandoned their study of numbers and invented Euclidean geometry after discovering that no number could measure the diagonal of a square. The thesis of this book supports a fascinating alternative that is both refreshing and plausible.

Grattan-Guinnes, Ivor. *From the Calculus to Set Theory, 1630–1910.* Princeton, New Jersey: Princeton University Press, 1980.
> The last part of this book, beginning with Chapter 5, is a deep history of the influence of Cantor's set theory on infinity, but it does require serious familiarity with reading mathematics.

Grattan-Guinnes, Ivor. *The Norton History of the Mathematical Sciences.* New York: Norton, 1998.
> This is a good resource book for history by a well-respected historian of mathematics. The first four chapters of this book are easily readable and cover the material in Part 1 of *Euclid in the Rainforest.* The second half requires a serious background in undergraduate college mathematics.

Lloyd, G. E. R. *Early Greek Science: Thales to Aristotle.* New York: Norton, 1970.
> This is a beautifully written, easily readable, small book that traces Greek science from the Pythagoreans through Plato to Aristotle.

Neugebauer, Otto. *The Exact Sciences in Antiquity.* Providence, Rhode Island: Brown University Press, 1957.
> A classic source for anyone interested in mathematics from Babylonian to Greek times. Neugebauer shows that the Babylonians had numerical tables approximating $\sqrt{2}$ and gives evidence that the Pythagorean theorem was known at least a thousand years before Pythagoras.

Proclus. *A Commentary on the First Book of Euclid's Elements.* Translated with introduction and notes by Glenn R. Morrow. Princeton, New Jersey: Princeton University Press, 1970.
> This is a remarkable translation of a fifth-century A.D. *Commentary on Euclid's Elements* and much of the mathematics predating Euclid. This is the book historians refer to for understanding the mathematics that led up to Euclid. We must remember that this book was written more than seven hundred years after the period it professes to know about, yet it remains one of the closest extant sources for such study.

Acknowledgments

Highest on my list of people to thank is my wife, Jennifer, who is both my greatest supporter and my most respected critic. To my brother, Barry, no words can say how much I love and admire him, but two words from him encouraged me to write this book. To Stephen Morrow at Pi Press, who took a chance with my jumble of notes and me to brilliantly organize and clarify my own ideas, and to foresee that they could turn into a book. To Jeffrey Galas, my editor, for his insightful suggestions and editing, which miraculously transformed what I said to what I meant to say. To Kristy Hart, my production editor, who, under pressure from time, graciously and expertly handled many changes.

I wish to express sincere appreciation to the many characters in this book responsible for countless pleasures of learning mathematics. In particular, to Roger Hooper, Jesus Lopez, Jeannot Berstel, and Uriah Brown, wherever they are. To Roger Godement, who might not recall that 40 years ago he patiently spent many hours with me on several occasions at the Café Luxembourg clarifying proofs I could not understand in his modern algebra classes, and paying for coffees.

Special thanks to David Port, Arlene Bouras, Peter Meredith, Laura Stevenson, and Franklin Reeve, who carefully read early drafts and gave excellent suggestions. To Tony Gengarelly, editor of *Mind's Eye*, who inspired me further by publishing a chapter. To Evan Johnson and Nick Sherefkin, for stimulating conversations about mathematics from the point of view of a teenager. To John Barnett, for wonderfully illustrating all the figures in this book in record time. To Sorina Eftim, for reading Part 3 and catching errors related to statistics and clarifying several points. To the Marlboro College library staff—Mary White, Elsa Anderson, and, in particular,

Radmila Ballada—for their gracious and expert assistance in finding cross-referenced material for accuracy checks. To Malcolm Brown and David Fowler, who, through many discussions lasting till dawn, introduced me to the philosophers and mathematicians behind Euclid and Greek mathematics. To Jim Markovitch, Iuliana Radu, Eric Weitzner, and Sylvie Weil. To The Café Beyond in Brattleboro, Vermont, for providing cappuccinos and a meeting place for many conversations surrounding the topics of this book. To Catherine Mazur Jefferies, Tom Jefferies, Tamina Clark, and Steven Clark. And, finally, I am grateful to my identical twin granddaughters, Sophia and Yelena Mazur Jefferies, for showing me how to learn new things.

Index

K

L

U